A History and Philosophy of Fluid Mechanics

G. A. TOKATY

Emeritus Professor of Aeronautics and Space Technology

THE CITY UNIVERSITY, LONDON

DOVER PUBLICATIONS, INC.

New York

Copyright

Published in Canada by General Publishing Company, Ltd., 30 Lesmill Road, Don Mills, Toronto, Ontario.

Published in the United Kingdom by Constable and Company, Ltd., 3 The Lanchesters, 162–164 Fulham Palace Road, London W6 9ER.

Bibliographical Note

This Dover edition, first published in 1994, is an unabridged, corrected republication of the work first published by G. T. Foulis & Co, Ltd, Henley-on-Thames, Oxfordshire, England, 1971, under the title *A History and Philosophy of Fluid-mechanics*. For the Dover edition the author has substituted a new Preface for the original one and a new photograph for Plate 15, which originally showed only Dimitri Pavlovich Riabouchinsky.

Library of Congress Cataloging-in-Publication Data

Tokaty, G. A.
[History and philosophy of fluidmechanics]
A history and philosophy of fluid mechanics / G.A. Tokaty.
p. cm.
Originally published: A history and philosophy of fluidmechanics. Henley on Thames : Foulis, 1971.
Includes index.
ISBN 0-486-68103-3
1. Fluid mechanics—History. I. Title.
QA901.T63 1994
532′.009—dc20 94-7411
CIP

Manufactured in the United States of America
Dover Publications, Inc., 31 East 2nd Street, Mineola, N.Y. 11501

Contents

Preface

The first edition of this book was sold out within two months. The initiative of the present (second) edition comes from Dover Publications, Inc., and I welcome it.

Except for small technical corrections, this book undergoes no changes. Its basic idea remains the same, and may be stated as follows:

The scientist should listen to every reasonable suggestion, but judge objectively. He should not be biased by appearances; have a favorite hypothesis; be of a fixed school of thought; or have a master in matters of knowledge. He should remember constantly that the progress of knowledge is often hampered by the tyrannical influence of dogma.

Anyone who has ever aimed at so noble a goal knows the great difficulty of such an attempt. But should this book achieve at least a portion of the author's ideal, he will feel happy.

G.A.T.
London

Basic definitions

Science is an ordered and systematic knowledge ascertained by theoretical analysis, observation and experiment. The science, or family of sciences, which aims to study and to master the material world in the interests of mankind, used to be (and still is, here and there) called Natural Philosophy. Its methods were and are: general logical deduction, scientific hypothesis, observation, experiment, comparison, analysis, synthesis, and so on. But in all these, in one degree and form or another, Mathematics plays a role: Natural Philosophy and Mathematics are as inseparable as a living cell and water.

There are, however, two kinds of Mathematics: Pure Mathematics, which studies abstract numbers, ratios, relationships, figures, functions, and works out and formulates abstract methods and theorems of mathematical philosophy and logic, and builds up a general mathematical culture, and Applied Mathematics, which is concerned with the study of physical, technological, biological, chemical and sociological worlds. To put it differently, a system utilizing, in addition to the purely mathematical concepts of space and number, the notions of time and matter, is Applied Mathematics. It includes the mechanics of rigid and deformable bodies, the theory of elasticity and plasticity, thermodynamics and biomathematics, statistics, etc. In a restricted sense, Applied Mathematics means the use of the concepts and theorems of Pure Mathematics for the study of the problems of Mechanics, which can be defined as the mathematical theory of the motions and tendencies to motion of particles and systems of particles under the influence of constraints, and the study of motions of masses and of the effects of forces in causing or modifying these motions.

The mathematical structure of Mechanics (i.e. its integro-differential forms) is largely due to Leonhard Euler (1707–83), Joseph Louis de Lagrange (1736–1813) and William Rowan Hamilton (1805–65); in this conception, it is often called Theoretical Mechanics, or Analytical Mechanics. One should not jump to the conclusion, however, that Theoretical (Analytical) Mechanics is something different or independent of Mechanics.

The French word *fluide*, and its English equivalent 'fluid', means 'that which flows', i.e. a substance whose particles can move about with complete freedom (ideal fluids) or restricted freedom (real fluids). That branch of Mechanics, or Applied Mathematics, which studies the laws of motion and tendencies to motion of fluids is called Fluidmechanics. When it deals, mainly or exclusively, with liquids, that is, when 'fluid' stands for 'liquid' (in most cases meaning 'water'), Fluidmechanics becomes the mechanics of liquids, or Hydromechanics.† When 'fluid' stands for 'gas' (in most cases meaning 'air'), Fluidmechanics becomes the mechanics of gases, or Aeromechanics.‡

Hydromechanics, in turn, may be subdivided into Hydrodynamics, Hydraulics, and Hydrostatics. The chief objective of Hydrodynamics§ is to establish theoretical-analytical relationships between the kinematic|| elements of motion, or flow, and forces which cause and maintain them. Hydraulics¶ studies the laws of motion of liquids in tubes, pipes, channels, elbows, and other engineering devices; as a general rule, its methods are based upon the basic concepts and theorems of hydrodynamics. Hydrostatics†† deals with the equilibrium of liquids at rest.

When 'fluid' stands for 'air' or '*aēr*' ('gas', more generally), Fluidmechanics becomes Aeromechanics. The latter may be subdivided into Aerostatics, Theoretical Aerodynamics, Experimental Aerodynamics, and Mechanics of Flight. Aerostatics‡‡ studies the laws of equilibrium of air and other gases at rest. Theoretical Aerodynamics§§ is the science of motion of gases in the Eulerian-Lagrangeian-Hamiltonian sense, i.e. uses the laws, theorems, axioms and general theoretical concepts of analytical (theoretical) mechanics for the study of air-gas flows under the action of forces; its role in Aeromechanics is similar to that of Hydrodynamics in the mechanics of liquids. The main objective of Theoretical Aerodynamics is to establish theoretical-analytical relationships between the dynamic, kinematic and thermodynamic characteristics of gas flows.

Experimental Aerodynamics verifies the theories, equations and formulae of Theoretical Aerodynamics by means of laboratory experiments and corrects them through the introduction of experimentally determined coefficients. In many cases, Experimental Aerodynamics succeeds where Theoretical Aerodynamics fails.

The Mechanics of Flight works out the general equations of flight and determines the air-speed and power required, climb performance and landing-take off characteristics, range and duration of flight, stability and

† *hydōr* = water + *mēchanikē* (the science of motion).
‡ *aēr* = air + *mēchanikē* (the science of motion).
§ *hydōr* = water + *dynamikos* (force).
|| *kinēmatos* = motion.
¶ *hydōr* = water + *aulos* (tube).
†† *hydōr* = water + *statikos* (equilibrium).
‡‡ *aēr* = air + *statikos* (equilibrium).
§§ *aēr* = air + *dynamikos* (force).

control criteria, and so on. In all these, it makes extensive use of the theories, theorems and fundamental concepts of almost all branches of Mechanics.

Finally, the fairly new notion, 'Aerospace', embraces and unifies Aeronautics and Astronautics, or Space Technology. Aeronautics† is the general name for all the aspects and problems of flight within the atmosphere of the earth. Astronautics‡ embraces all the aspects and problems of flight beyond the atmosphere, and may be defined as the science of motion of rockets, sputniks and spaceships beyond the atmosphere.

Fluids and life

How important are fluids in life? Here are several examples. Human beings, animals and vegetation are literally water-based. Every living cell in your body has a fluid interior, a vital solution of various substances in water; human blood is more than nine-tenths water; our muscles average 92 per cent water; all in all, man's body contains about 71 per cent water by weight – and this water, evaporating and flowing from the surface of the body, breathed out as vapour in breath, must be continuously replenished if the body is to remain alive. Man pours down his throat five times his weight in water every year; by the time he dies, if his life span is normal, he will have drunk about 6500 gallons of it.§

Consider the matter also from another angle. The desert is arid. The sun, the other source of life, kills almost every living thing there. In just one hour, it fills each square yard of the desert with more than 800 large calories – a large calorie is the amount of heat required to raise the temperature of 2·2 pounds of water by one degree centigrade. An egg can be fried on the burning sand. The Kara-Kum desert in Central Asia (USSR) covers almost 150000 square miles. Its surface receives a stupendous amount of solar energy. But fresh water is scarce, if not absent, although whole seas of salt water lie just below thirsty pastures. Flocks of hardy sheep (three million head) roam the desert. Their numbers could increase were it not for the shortage of fresh water. It is easy to imagine the economic benefits of irrigating the desert.

It has been estimated that a mere 1·3 cubic yards of water is needed to produce 8 pounds of wool, nearly 17 pints of milk, up to 22 pounds of

† *aēr* = air + *nautes* (navigation).
‡ *astron* = star + *nautes* (navigation).
§ *Water – the mirror of science*, Kenneth S. Davis, John A. Day, Anchor Books, Garden City, New York: 1961.

meat, or to grow 4 ounces of cotton.† No wonder that the Soviet Union‡ has decided to harness this solar energy, and a helio-installation to distil salt water is now being built in the heart of the desert. The installation will be a sizeable complex comprising distillers (each over 700 square yards in size), ferro-concrete reservoirs able to hold 740 cubic yards of water, a well, water-raising pipes, solar electric generators, water troughs, etc.

A desperate effort, indeed; but a necessary one. Perhaps efforts of this kind will convince man that water and air are far too fundamental for all forms of life for them to be treated too barbarically. It seems incredible, for example, that, in the same Soviet Union, millions and millions of cubic metres of waste water are daily being dumped into fresh sources, making them almost useless. Even in the USSR, which has a planned economy and therefore cannot be regarded as the most wasteful country in the world, each year some 3 000 000 tons of acids, 2 000 000 tons of oil products, 1 000 000 tons of fats, and hundreds of thousands of tons of salt, fibres and metal are dumped into rivers.

In the United States (undoubtedly the most wasteful country) whole rivers and lakes are being polluted to such an extent that they can no longer be used for drinking or swimming purposes. The annual American waste output (of which a large part goes into rivers and lakes) includes something like 142 million tons of smoke and fumes, 7 million discarded cars, 20 million tons of waste paper, 48 billion tin cans, 26 billion bottles and jars, 3 billion tons of waste rock and milling swarf, and 50 trillion gallons of hot water.

But man spoils not only the waters of the earth. You may have heard of the notorious 'smog' in Los Angeles and London. But in actual fact it exists in every industrial centre. Combustion of fossil fuels has increased carbon dioxide in the atmosphere by one-tenth in a century, and may attain an increase of a quarter by the year 2000: this would be catastrophic for weather and climate, and seriously damaging to every form of life.

When analysed carefully, this 'Battle of Fluids', the struggle for their protection and destruction, reveals that it, too, is based upon the laws of Fluidmechanics. We may, therefore, assume that knowledge of the history of the subject is important also from the point of view of learning how to minimize the damage to the atmosphere and water resources, if not how to prevent them from being damaged at all. But to be able to do this, that is to understand what exactly is damaged or protected, it is necessary to know, first and foremost, the compositions of the fluids themselves.

† *Sputnik*. June 1969: Moscow.

‡ For 'Soviet Union', which, of course, no longer exists as such, read 'ex–Soviet Union', and for 'USSR' read 'ex-USSR'. (1994 note.)

Water and air

The composition of water has been the subject of intensive study since the emergence of chemistry and physics. But it seems to be impossible to say who exactly started it and when. The only thing we know for certain is that (already) Hero of Alexandria was anxious to know why water boiled and produced steam. We shall discuss this period of fluidmechanics later on.

E. W. Morley, of Western Reserve University, Cleveland, Ohio, in 1895, reported that the weight of Oxygen to Hydrogen in water was as 7·9395 to 1·00000, and that the volume ratio was as 1·00000 to 2·00288. F. P. Burt and E. C. Edgar, of England, in 1916, considered, on the basis of their experiments, 7·9387 to 1·00000 as the most exact weight ratio. The present value accepted by the International Union of Chemistry, Committee of Atomic Weights, is 15·9994/2 to 1·00797.

Water can be in two major states: solid (ice) and liquid (water). The first of these occurs at the so-called 'freezing point'. In other words, at a pressure of one atmosphere, ice melts (becomes water) at 0°C. If the water is pure, especially when there are in it no dissolved gases, it can be heated without boiling up to 100°C and even higher. But, in normal conditions, 100°C is the 'boiling point', at which temperature steam formation is so intense that it occupies a volume 1 700 times greater than the water itself.

By 'one atmosphere' is meant the pressure exerted by the atmosphere as a consequence of the gravitational attraction exerted upon the column of air lying directly above the point where the pressure is measured. One atmosphere = 760 mm of Hg. When the pressure is 770 mm, the boiling point occurs at 100·366°C, when it is 750 mm at 99·360°C, when it is 740 mm at 99·255°C, when it is 730 mm at 98·877°C, when it is 388 mm at 81·7°C, when it is 76 mm at 46·1°C, when it is 1 520 mm at 120·6°C, when it is 7600 mm at 180·5C. Hg stands, of course, for the Latin name *Hydrargyrum* or Mercury (the element): atomic weight 200·59, atomic number 80, melting point —38·87°C, boiling point 356·58°C. Mercury is a silver-white liquid metal, the only metal that is liquid at ordinary temperatures. It is widely used in Fluidmechanic measuring instruments (manometers, for instance) because of its remarkable properties; to name just one of them, it does not wet glass, which is so important in manometry.

Among the other physical characteristics of water, and of fluids generally, the most important one is density. The amount of mass per unit volume is the mass density ρ; the weight per unit volume is the weight density γ. The two are connected by the equation $\gamma = \rho g$, where g is the gravitational acceleration. Both ρ and γ are different for different fluids and depend on the temperature, as shown in the table below:

Liquid	Temperature	γ kg/M^3	ρ kg. sec^2/M^4
Water	15	999	102
Sea water	15	1 020	104
Mercury	15	13 595	1 385
Castor oil	15	970	99
Paraffin	15	790–820	81–84
Benzine	15	680–720	69–73
Acetone	20	790	80·5
Benzol	0	900	91·9
Beer	—	1 040	106
Chloroform	18	1 480	151
Methyl alcohol	0	800	81·5
Alcohol	15	790	80·5
Petroleum	19	760–900	78–92

Natural waters may be contaminated with insoluble suspended materials, soluble inorganic matter, soluble organic matter. In oceans, seas and salt lakes, water's principal content is sodium chloride, with small amounts of calcium, magnesium, potassium, and sulphate, carbonate, and many other elements in smaller concentrations.

And now about Air. By this is meant the whole mass of the earth's atmosphere, whose average percentage composition may be represented by the following table:

Nitrogen	78·08
Oxygen	20·95
Argon	0·93
Carbon dioxide	0·03,

and some other gases. All the figures are given for the so-called dry air, i.e. for air from which all water vapour has been removed.

Although air is composed of various gases, in most cases it is regarded as a uniform gas. How do the main physical characteristics of such a gas compare with other gases? The table below gives the answer:

Gas	Pressure	γ kg/M^3	ρ kg. sec^2/M^4
Air	760 mm	1·188	0·121
Oxygen	760 mm	1·312	0·134
Hydrogen	760 mm	0·0827	0·00813
Helium	760 mm	0·164	0·0167
Nitrogen	760 mm	1·151	0·1174

It is a matter of ordinary daily observation that all fluids are capable of exerting pressure. A certain amount of effort is necessary in order to immerse your hand in water and to move it there. The effort is much less when you do the same in the atmosphere, in a gas generally, because its density is much smaller. That the atmosphere at rest exerts pressure is shown directly by means of an air pump. Amongst many experiments, a simple one is to exhaust the air within a receiver made of very thin glass; when the exhaustion has reached a certain point depending on the strength of the glass, the receiver will be shivered by the pressure of the external air. The action of wind, the motion of a wind mill, the propulsion of a boat by means of sails, and other familiar facts offer themselves naturally as instances of the pressure of the air when in motion.

All such substances as water, oil, mercury, steam, air, or any kind of gas are fluids, but in order to obtain a definition of a specific fluid, we have to find a property which is common to all these different kinds of matter, and which does not depend upon any of the characteristics by which they are distinguished from each other. This property is found in the extreme mobility of their particles and in the ease with which these particles can be separated from the mass of fluid and from each other. In other words, a fluid is a substance whose particles can be very easily separated from the whole mass.

Gases (air, for instance), by the application of ordinary force, can be easily compressed, and, if the compressing force be removed or diminished, will expand in volume. Liquids, too, are really compressible, but experiments by Canton in 1761, Perkins in 1819, Oersted in 1823, Colladon and Sturm in 1829, and others, have proved that they are only slightly compressible.†

The pressure of a liquid at rest is entirely due to its weight, and to the application of some external force, while the pressure of a gas, although modified by the action of gravity, depends in chief upon its volume and temperature. The action of a common syringe will serve to illustrate the elasticity of the atmospheric air. If the syringe be drawn out and its open

† *Elementary Hydrostatics*. W. H. Besant, fifth edition, Cambridge: 1873.

end then closed, a considerable effort will be required to force in the piston to more than a small fraction of the length of its range, and if the syringe be air-tight, and strong enough, it will require the application of very great power to force down the piston through nearly the whole of its range. And this experiment was good enough for our ancestors to prove that the pressure of the air increases with the compression, the air within the syringe acting as an elastic cushion.

They also found that if the temperature is increased, the elastic force of a quantity of air or gas which cannot change its volume is increased, but that if the air can expand, while its pressure remains the same, its volume will be increased. To illustrate this, early researchers used to employ an air-tight piston in a vertical cylinder containing air, and let it be in equilibrium, the weight of the piston being supported by the cushion of air beneath; then they used to heat it and thereby to raise the temperature of the air in the cylinder: the piston rose, or, if not allowed to rise, the force required to keep it down increased with the increase of temperature.

So, simple experiments provided information of fundamental importance. Man began to know the physical laws governing the thermal and elastic properties of the atmosphere. In due course, it became common knowledge that the mass density, weight density, pressure, temperature and chemical composition of the atmosphere change with altitude: we shall discuss these matters later on, in connection with Pascal's and others contributions to Fluidmechanics.

The first uses of fluids

To tell the whole story of the role of fluids in the growth of physical sciences, we ought to go back to the dim ages of the emergence of organized life, to the times before our cave-dwelling ancestry began to wonder why water ran downhill. But this we cannot do in a book of this size and purpose. We must, however, glance at the patterns of very early 'river civilizations', because we have in them interesting evidence of the uses of fluids in the interest of man, and of life. A large mass of evidence suggests that, sometime before 5000 BC, a peaceful, artistic and highly talented race left their homes somewhere in Central Asia and descended into Mesopotamia, 'the land between the rivers', Euphrates and Tigris, which sometimes is called the cradle of human civilization.†

† *The Growth of Physical Science.* Sir James Jeans, Cambridge: 1947.

They had considerable engineering skill, as is shown by their irrigation methods and techniques as well as by their temples and palaces.

The astronomical knowledge of these civilizations reached the Egyptians, who used it to determine their civil year to be exactly 365 days, 12 months of 30 days each, together with 5 extra sacred or 'heavenly' days. The day on which the star Sothis (our Sirius) first became visible was found to coincide very approximately with the Nile floods, and so the state of the great river became a sort of calendar. Furthermore, the Egyptians kept a record of successive risings of the Nile and analysed the behaviour of flood waters. This led them to one of the most interesting inferences, that there existed in nature interaction between the seasons, air and water. It may be relevant to mention here also that sometime around the year of 3000 BC, the Chinese, too, thought that there existed a magic connection between winds, clouds, rain and river floodings.

More generally, the systematic study of the materials of antiquity provides adequate evidence that, indeed, scientific knowledge, as almost anything else related to early civilization, dawned in the East. India, China, Mesopotamia-Babylonia, Egypt, Phoenicia and other civilizations gave birth to many aspects of organized life, and they were the first beginners in the field of fluids, in the study (although, perhaps, 'study' is too strong a word here) of the natural laws of the atmosphere, climate, rivers, and so on. We may say that those who suffered most must have learnt from their sufferings more than others. We can be sure that those who experienced the devastating attacks of tornadoes could not fail to notice their peculiarities and that those who had to go through extremely hot and humid summers could not be ignorant of the vast changes in the physical state of the atmosphere.

Life was so full of blisters, it offered so many miseries, its lessons of humility were so varied and so numerous, that the Egyptians and others had no alternative but to lead it in the hard way; which made them a very knowledgeable and experienced nation. As Francis Bacon (1561–1626), the English philosopher and author of a number of important theories, wrote:† the man, being the servant and interpreter of nature, can do and understand so much and so much only as he has observed in fact or in thought of the course of nature. The Egyptians knew a great deal about fluids. Their knowledge by experience and their human power met in one. Their civilization, probably one of the more advanced civilizations of antiquity, emerged in close association with what was going on on the

† *The Philosophers of Science*. Edited by Saxe Commins and Robert N. Linscott, The Pocket Library, New York: 1954.

Nile banks.† Probably earlier than anybody else, they knew that fluids, water, and air, not only destroy, but, if used wisely, can help man immensely. The very fact that the Egyptian civilization developed along the banks of their great river Nile is more than suggestive: it indicates that already then, 3000–600 BC, civilization and water were inseparable.

The Egyptians saw how, almost every year, the great river flooded its banks, and how it washed away their cultivated lands, homes, roads, etc. Each new disaster of the kind was a new lesson, for them. They began to discover, by experience, that wherever the cross-section of a flow tended to get narrower, the flow became faster and deeper; and wherever this occurred, the danger was real. So, without knowing anything about Leonardo da Vinci's velocity-area law (which will be discussed in due course), the Egyptians began judging the dangers ahead on the basis of the same law.

One of the outstanding characteristics of the Egyptian civilization was its relative fairness towards the loser. Each time the waters of the Nile wiped out areas of the fertile land, the remaining land was redistributed anew. This required certain methods and techniques of measuring and re-measuring. And it was, perhaps, this which brought to life the famous 'Egyptian Triangle': a rope of twelve units in length divided by knots into sections of three, four and five units, and then made into a triangle, with the knots at the corners. The angle opposite the side of five units is always a right angle (Figure 1). The triangle could be made of any size, which made it a convenient device to measure out the land in such triangles and rectangles.

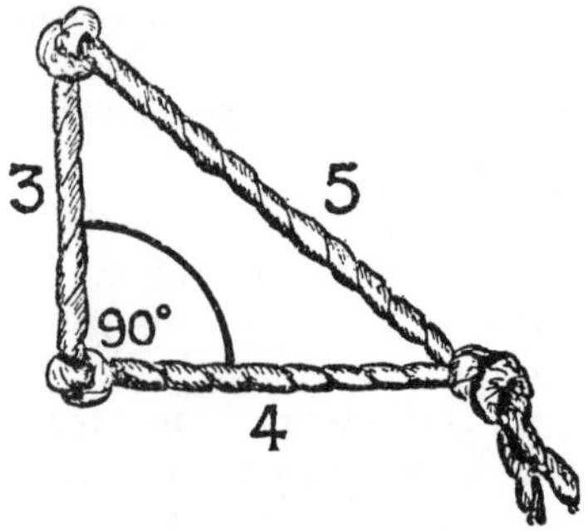

Fig. 1. The Egyptian Triangle: a rope of twelve units in length divided by knots into sections of three, four and five units, and then made into a triangle, with the knots at the corners. The angle opposite the side of five units is always at right angle

† *Ancient History of Egypt, Assyria and Babylonia.* Elizabeth M. Sewell, Longmans, Green, London: 1845.

This triangle embodies the well-known Pythagorean Theorem (which says that the sum of the squares of the lengths of the sides of a right-angled triangle is equal to the square of the hypotenuse). But, strictly speaking, the real mathematical author of the theorem was Euclid, or Euclides (about 300 BC), a Greek geometer, founder of a school in Alexandria, whose basic work *Elements* (13 books) became the basis of Geometry.†

The fact that the Egyptians built their great Pyramids, or Tombs (in which they buried their kings and other dignitaries), so successfully that they lasted up to 4000 years, also suggests that at least some simple laws of fluidstatics and fluidmechanics must have been known to them, probably not mathematically but as a matter of life experience. As an old Ossetian tale asserts, *dony ahdau Pyrsyie arbacyd*, water's behaviour originates from Egypt. Perhaps the sufferers from the Nile floods knew that if water can wipe out whole areas, it can also do useful jobs for man; perhaps they knew that, once elevated somehow, water possessed lots of energy almost readily available to man. Is it too much to suggest that a civilization that existed for so long a time along so powerful a river could not be unaware of canal building, irrigation principles, flow deflectors,

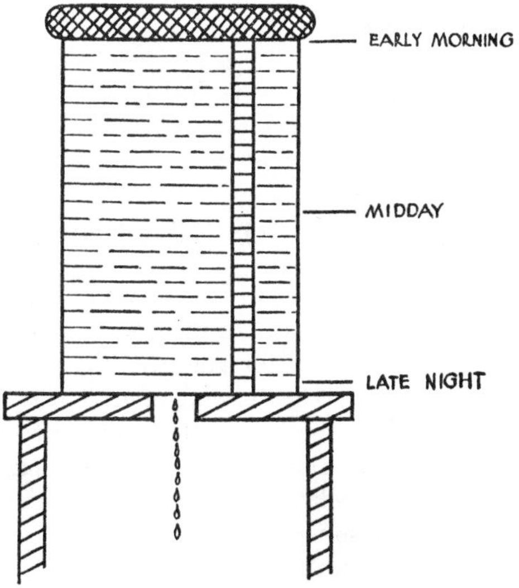

Fig. 2. The Egyptian Water Clock: water was allowed to drop regularly from a large container through a small hole; the lowering of the free surface was calibrated in relation to the movement of the sun

† *Euclid's Elements* (translated by T. L. Heath), 3 volumes, Cambridge: 1908.

bridges, and simple water elevating devices? After all, history tells us that the Egyptians of the Nile valley built themselves boats and ships, in which they used to cross the river and even to venture down into the Mediterranean.

Then there was the so-called Egyptian Water Clock, or *Clepsydra*, which worked as follows: water was allowed to drop regularly from a large container, a stone vessel of a definite configuration (truncated cone, for example), through a small hole (Figure 2); the downward movement of the free surface of water in the vessel was calibrated in relation to the movement of the sun, thus showing time. Of course, this was still far from being applied fluidmechanics in the modern sense of the word, but when you analyse the device thoroughly, it becomes clear that its inventors must have known, from experience, a great deal about the fluid properties of water.

The Babylonian civilization in south-west Asia was roughly of the same age and character as the Nile civilization. But the Babylonian lands were less fertile, the climate was less generous; therefore, to feed themselves, the population had to water its Alma Mater, the land, much more vigorously; which demanded irrigation systems, if (*ad hoc*) systems is not too strong a word; and any irrigation involves many elements of what we call today hydraulics.

Mythology and fluids

We should, however, not jump to the inference that man's attitude to fluids has always been rational. On the contrary, in very many cases, even in Egypt and Babylonia, natural disasters like floods, torrential rains, storms, etc., were thought to be the work of mysterious forces. Moreover, water and air themselves, too, were often considered to be mysterious. The don't-touch-it state of mind revealed that what men and women could see for themselves, what they drank and breathed daily, was nothing less or simpler than supernatural.

The amazing tales of the Myths of Greece and Rome had no room for conjecture: a hand mighty enough to call all the wonders of life into being could also create air, water, rivers, lakes, seas, oceans, rain, etc. These myths gave rise to others. Suppositions became certainties. And soon myths and fables evolved to be handed down from generation to generation. At first, land, sea and air were mixed up together, so that

Earth was not solid, the sea was not fluid, the air was not transparent. As Ovid put it,†

> No sun yet beam'd from yon cerulean height;
> No orbing moon repair'd her horns of light;
> No earth, self-poised, or liquid ether hung;
> No sea its world-enclasping waters flung;
> Earth was half air, half sea, an embryo heap;
> Nor earth was fix'd, nor fluid was the deep;
> Dark was the void of air . . .

Over this shapeless mass reigned a carefree deity called Chaos, whose personal appearance could not be described, as there was no light. He shared his throne with his wife, the dark goddess of night, named Nyx or Nox. Their son, Erebus, in due course married his own mother Nyx. So Erebus and Nyx ruled over the chaotic world until their two children, Ather (Light) and Hemera (Day), acting in concert, dethroned them and seized supreme power.

Space illumined for the first time by their radiance, revealed itself in all its uncouthness. Ather and Hemera decided to evolve from it a thing of beauty. They summoned their son Eros (Love) to their aid. By their combined effort, the sea and the earth were created. In the beginning the earth was ugly. Eros did not like this, therefore he seized his life-giving arrows and pierced the cold bosom of the earth. Immediately the brown surface was covered with luxuriant verdure; birds of many colours flitted through the foliage of the new-born forest trees; animals of all kinds gambolled over the grassy plains; and swift-darting fishes swam in the limpid streams.

In short, there was now everything: earth, water, air, light, life. But what was the shape of the earth?' A flat thing, of course. And at its ends? The great river Oceanus in a 'steady, equable current', from which the streams and rivers of Greece derived their waters.

The myths of the sea comprise, of course, Oceanus and Neptune (the earth shaker). We are informed that Neptune's place was beneath the deep waters near Greece. Neureus, another personification of the sea, is quite inseparable too from his native element, as are also the Tritons, Oceanides, Nereides, and the alluring Sirens who, however, have also been viewed as personifications of the winds.

The cloud myths comprise not only the cattle of the sun, the Centaurs, Nephele, Phryxus, Helle and Pegasus, but the sky itself was a blue sea, and the clouds were ships sailing over it.

† *The Myths of Greece and Rome*. H. A. Guerber, George G. Harrap & Company, London: 1910.

In the myths of the wind, Mercury was one of the principal personifications. In addition to his so many other qualities, he was thought to be the 'lying, tricksome wind god who invented music', for his music was but 'the melody of the winds, which can awaken feelings of joy and sorrow, of regret and yearning, of fear and hope, of vehement gladness and utter despair'. Mars, too, was a personification of the wind. His nature is further revealed by his inconstancy and capriciousness; and whenever he is overcome, he is noted for his great roar. His name comes from the same root as Maruts, the Indian god, and means the grinder or crusher. It was first applied 'to the storms which throw heaven and earth into confusion', a clear reference to storms and thunderstorms.

It is interesting to mention that Otus and Ephialtes, the gigantic sons of Neptune, were also at first personifications of the wind and hurricanes.

And so on, and so forth. Obviously, the ancient Greek mythologists, and, may I add, also the ancient mythologists of the Caucasus† were very much concerned with what was going on in nature, but were unable to do more than compose myths and tales, which played, however, a role in the primordial formation of the history and philosophy of the science of fluids.

Plato and fluids

Life has always been water, water has always been life, but man, the main hero of life, its principal actor, who needed it so much, never hesitated to spoil it. Almost all the great enlighteners glorified the life-supporting fluid, but so many others destroyed it so mercilessly. As Antoine de Saint-Exupéry, the distinguished French writer, put it, 'O, Water, you have neither taste nor colour or odour, you are indescribable, yet you give so much pleasure and delight, without anybody knowing what you are! One cannot say that you are necessary in life – you are life itself. You fill us with joy, which cannot be explained by our feelings. . . . You are the greatest wealth in the world.'

But what *was* water, air? Thales of Miletus (624–546 BC), one of the earliest and probably the greatest (for his time) Greek philosopher, wanted to know the answers. Moreover, he was anxious to find out the general configurations of the major fluids, their volume forms. And he came to the wrong conclusion that all things in the material world were

† See, for example: *Iron Adamy Arhautta*, a book in Ossetian, by S. A. Britaev and G. Z. Kaloev, 435 pages, Ordzonikidze (now Vladikavkaz), 1960.

originally produced from water; that it was, therefore, superior in relation to anything else ('water is best'); and that, accordingly, man had the duty of 'respecting and loving water as the source of existence'. For, indeed, he wrote, that from which everything is made must always be considered as the first principle.

There is no need to agree or disagree with these concepts and assertions. We simply accept the fact that Thales was among the very first thinkers who tried to create a natural philosophy of fluids, whose approach constituted a significant departure from the myths and tales of seas, rivers, streams and air: that in itself should be accepted as a valuable contribution to the future science of fluids.

Pascal once remarked that things are always at their best in their beginning. But this was not so in the history of Fluidmechanics. Almost all the beginners, some of them great figures in history, asserted that there were in the universe only four basic elements: earth, water, air, and fire. They, these elements, were to be found in spherical configurations, all around one centre. The hard core of the system, the earth, was by far the heaviest element, therefore it occupied a 'bottom position'. The next heaviest element was water: immediately above the surface of the earth. Then air, then fire.

Among the protagonists of this four elements theory were Pythagoras (probably 580–500 BC), Empredocl or Empredokl or, usually in English, Empedocles (490–430 BC), Plato (427–347 BC), Aristotle (384–322 BC), and others. The theory was 'finalized' by Dante (1265–1321).

Plato's views are of particular interest to us; he taught that each of the elements existed, in one form or another *before* the 'creation of the world by God', thus clearly siding with the philosophy of Heraclitus (530–470 BC), who asserted that 'the world, all in one, was *not* created by any God or any man, but was, is and will ever be a living flame, systematically flaring up and systematically dying down'.† Frederick Engels (1820–95) probably the greatest of all the dialectical materialist-philosophers, made the point even clearer: the materialistic outlook on nature, he wrote in his *Ludwig Feuerbach*, means no more than simply conceiving nature just as it exists, without any foreign admixture.

But to return to Plato: in his metaphysical theory, he attempted to answer the question, *What is real?* He looked at a stream of water and asked: is what I see real? Winds, storms, rain, floods, etc., are these real? And he denied uncompromisingly the reality of what we call the real world, the objects and natural phenomena we see and observe. It should, however, be pointed out that this rejection of the sensible world as not real is no special peculiarity of Plato's; it is a position that is shared, to

† [*On the Nature of Things*], Lucretius, George Bell and Sons, London: 1886.

some degree, by almost all the leading thinkers of ancient Greece. Furthermore, we cannot say that Plato was a fool. Indeed, when one studies his philosophies, it becomes clear that his rejection of reality was directed towards the discovery of what *is* real. When he declared, for example, that *παντελῶζ ὄν, παντελῶζ γνωστον*, the completely real is the completely intelligible, he meant to say that things and phenomena may certainly be real, if they are expressible either verbally or otherwise.† Pythagoras and others made use of this general formula and arrived at the conclusion that any natural fact and/or phenomenon must be expressible in mathematical terms.

On the other hand, Plato thought that all material bodies are made up of the four elements, but in different bodies these elements (earth, water, air, fire) are combined in various ways and various proportions. The elements in turn differ from each other by being composed of minute particles differing in their geometrical shapes; and these geometrical shapes are again made up of different combinations of simpler geometrical figures.‡ Now, the modern physicist will tell us that this was certainly a scientific way of thinking, which, by its implications, is far from a blind rejection of reality.

Two different narratives concerning the destruction of the first humanity have reached the second: the Babylonian, repeated in the Book of Genesis, about the flood, and the Egyptian, written by Plato, about Atlantis. For twenty-five centuries human beings have been trying to solve Plato's riddle. His interest in fluids assumed dramatic forms, but we still face the same question: what is Atlantis, myth or history? 'He who created it, also destroyed it', laughs Aristotle.§ This means, Atlantis is a myth. But to say this, Aristotle must have been a transcendently ruthless enemy of Plato; perhaps he was.

This, however, does not solve the problem: Plato created more riddles and problems than he ever resolved. To a scientist, there can be only one explanation: Plato was a non-scientist who tried to develop (among other things) scientific philosophies. To him a philosopher was a man possessed of the power of scientific thought without being a scientist; admittedly, the very notion 'science' was still new in the general reason, it was in words and arguments rather than in terms of mathematics and physics. But the age of scientific knowledge requires scientific facts and proofs and hence the understandable difficulties with Plato.

† See, for example: *Masters of Political Thought*. Michael B. Foster, vol. 1, George G. Harrap and Company, London: 1942.

‡ *The Philosophy of Plato*. G. C. Field, Oxford University Press, London: 1949.

§ *The Secret of the West*. D. Merezhkovsky, Jonathan Cape, London: 1933.

Aristotle and the science of fluids

Aristotle was born at Stagira, a Greek colony on the Macedonian coast, in the year 384 BC. In 367 BC he migrated to Athens in order to study philosophy under Plato, and for 20 years remained in the Academia (of Plato). In 342 BC, Aristotle became a private teacher to Prince Alexander (later called 'the Great') of Macedonia. In 334 BC, he returned to Athens and opened a school of philosophy, and this was about the time he became a very well-known figure.

There is probably no parallel in the history of mankind to the influence of this one man, either in the vastness of his intellectual achievement or in the extent of his influence extending over two thousand years, on the growth of scientific knowledge. I should, however, deal only with his contributions to the subject of this book. Many of his assumptions and theories proved to be wrong, but he created an intense scientific curiosity, which accelerated the emergence of sciences. He failed to penetrate the physical inner structure of the world, but he succeeded in the formation of certain basic laws of nature whose value to mechanics generally and fluidmechanics in particular cannot be doubted.

One particular aspect of Aristotle's writings influenced scientific thought negatively for many centuries after his death in 322 BC: his rejection of the concept of vacuum. He made truly heroic efforts to prove that nature does not tolerate 'nothingness' (nature abhors a vacuum): a perfectly valid theory which will be elaborated in this book later on; but Aristotle's reasons behind this theory were quite inaccurate.

The range of his interests and contributions was immense. Among other things, he stated,† quite categorically, that since every body in the universe has a 'heaviness', it must, and does, tend to move (to fall) towards its 'natural place', therefore the lower layers of matter, for example of water and of the atmosphere, must be more dense than the upper layers: a perfectly accurate conclusion of great value to the future fluidmechanics. From this theory of 'heaviness' he derived also the revolutionary conclusion that the atmosphere has a spherical configuration around the earth. But he was wrong when asserted that if a weight falls from a given height in a given time, twice that weight will fall from the same height in half the time. It was not until one thousand nine

† *On the Heavens*. Aristotle (English translation by W. K. Guthrie), William Heinemann, Ltd., London: 1936.

hundred years later that Galileo showed experimentally that Aristotle was wrong, and that, in actual fact, bodies of all weights fall from the same height in the same time.

All branches of Fluidmechanics rest upon the continuity principle. In brief, this is the principle that mass is indestructible and may be completely accounted for at different points of any fluid, at rest or in steady motion. And Aristotle was the first to give its general formulation. 'The continuous may be defined', he wrote,† 'as that which is divisible into parts which are themselves divisible to infinity, as a body which is divisible in all ways. Magnitude divisible in one direction is a line, in three directions a body. Being divisible in three directions, a body is divisible in all directions. And magnitudes which are divisible in this fashion are continuous.'

What a remarkable definition for the Aristotelian times! And that was not all. Like his predecessors and contemporaries, Aristotle believed that the 'four elements' theory of the universe was a true theory. But he made a significant step forward. Each of the elements, he asserted, was nothing more than the combination of different properties of one and the same thing – matter, *materia*. Water was so 'wet and moist and heavy', the earth so 'heavy and hard', the air so 'light and tangible', and the fire so 'hot', that their existence could not be doubted: they were real, therefore the whole universe was real.

But each of the elements was not an element in itself – it was a combination of material properties. Something dry combined with something hot produced fire; something cold combined with something dry gives earth; something moist combined with something cold gives water; and so on. In this way, according to Aristotle, the elements could transform into each other and thus exchange their properties. Between the full transformations there were the various physical bodies of the world: stone, iron, glass, vegetation, rivers, rain, etc.

We, the aerodynamicists, are obliged to Aristotle for his pioneering concepts dealing with the motions of projectiles and air resistance.‡ It was he who pointed out, for the first time, that when a body moves in the atmosphere, the surrounding air becomes hot and, in certain circumstances, it (the body) even melts – yes, he said, 'melts'. Of course, he was not an aerodynamicist; of course, he knew nothing of the modern mathematical theories of aerodynamic stagnation temperatures, kinetic heating, and the like: which makes his general ideas even more exciting.

Modern science and technology are indebted to Aristotle also for his more general ideas about motion. In his *On the Heavens*, he developed,

† *On the Heavens.*

for example, a distinction between the so-called *natural motion* and *enforced motion*. 'If you set a stone free, it will move (fall) freely naturally, in a definite direction. But if you force it to change this direction by means of applying to it a force, you will have an enforced or unnatural motion'. This conception led him to the conclusion that

> It is impossible to say why a body that has been set in motion in vacuum should ever come to rest. Why, indeed, should it come to rest at one place rather than at another? As a consequence, it will either necessarily stay at rest or, if in motion, will move indefinitely unless some obstacle comes into collision with it.

To understand and appreciate the historic importance of this formulation, we should perhaps recall that, in his *Mechanique*, Galileo (1564–1642), too, wrote that a body is 'indifferent to motion or to rest, and does not itself show any tendency to move in any direction or any resistance to being set in motion'. Huyghens (1629–95) in his *De motu corporum ex percussione*, wrote that 'any body in motion tends to move in a straight line with the same velocity as long as it does not meet an obstacle'. And Sir Isaac Newton (1642–1727) in his *Philosophiae naturalis principia mathematica*, concluded that 'every body perseveres in its state of rest, or uniform motion in a straight line, unless it is compelled to change that state by force impressed thereon.'

We thus see that Aristotle was, really, the first, Galileo the second, Huyghens the third, and Newton only the fourth milestone in the history and philosophy of the law of inertia. This alone gives Aristotle a prominent place among the first fathers of general Mechanics and of Fluid-mechanics. But he managed to go even further, although not always in the right direction. He disliked, for instance, the idea of absolute vacuum. His epistemological mind seemed to have worked roughly as follows: the most dense layers of the atmosphere are those nearest the surface of the earth; as the altitude increases, the density decreases; if we accept this natural law, then there must be an altitude beyond which there is no density, i.e. no mass at all, an absolute vacuum or nothingness; but what, then, occupies the space?; how, for instance, can light rays go through nothingness?

'We maintain, therefore', he wrote,† 'that fire, water, air and earth are transformable one into another, and that each is potentially latent in the others. But neither air nor fire fills the space between the earth and outermost heavens. The space between the heavenly bodies must be filled with some substance, which is different from air and from fire, but which

† *Meteorologica* 1, III. Aristotle (English translation by H. D. Lee), William Heinemann, Ltd., London: 1952.

varies in purity and freedom from admixture, and is not uniform in quality, especially when it borders on the air and the terrestial region. . . .'

What a fascinating conclusion it must have been for those times! And not only for those times. Try to recall some of your own school experiments. In a glass vessel, closed to the air on all sides, was an electric bell, compelled to ring continuously by an electric current, the source of which was also inside the glass vessel. Then, by removing the air from the vessel by means of an air-pump, you became distinctly aware of the fact that the sound of the bell was growing fainter and fainter, until it finally died away completely. You were thus taught to conclude that sound consists of a succession of compressions and rarefications of the air, and that, consequently, a sound wave cannot come into being in a space devoid of air, i.e. in a vacuum.

Then there was, or could be, another conclusion. The visibility of a body depends on the fact that rays of light proceeding from it are reaching our eyes. In a completely dark room we are unable to see anything. Only when we admit light, and its rays are reflected by the objects, do these objects become visible to us. But in your glass vessel, with its rarefied air, when you could no longer hear the ringing of the bell, you saw all the time how the hammer was knocking against the bell. Consequently, rays of light must have been capable of being propagated without air, i.e. capable of going through a vacuum. You may then say that when you see the sun and the stars in the sky, this, too, means that their light is able to rush through the empty space of the universe.†

But how does one conceive this? Aristotle and his able followers appealed to the 'world-aether' or simply 'aether' concept, without which the propagation of light would remain an unsolved riddle. They had to resort to this old child of sorrow in theoretical physics, because there were no Quantum Mechanics, Nuclear Physics, or general Field Theory. A refuge from the horror of vacuum was found, the general concept of universal continuity of matter and energy (then still unknown) was safeguarded – at least for the time being.

Just one more interesting point. Aristotle was a good observer. His eye noticed and his brain considered the fact that water or any other liquid never has an inclined surface. 'Why is this?' he asked.‡ 'Because if heaven (1) revolves in a circle and (2) moves faster than anything else then it must be spherical. If water is found around the earth, air around water, and fire around air, the upper bodies will follow the same pattern . . .' And this is how he proved his proposition in regard to water:

Let $\beta\epsilon\gamma$ be an arc of a circle, whose centre is α (Figure 3). Then the

† *Relativity and the Universe.* Dr Harry Schmidt, Methuen & Co. Ltd., London: 1921.
‡ *On the Heavens*, II, IV, p. 161.

line $\alpha\delta$ is the shortest distance from α to $\beta\gamma$. Water will run towards δ from all sides until its surface becomes equidistant from the centre. It therefore follows that the water takes up the same length on all the lines radiating from the centre; and remains in equilibrium. But the locus of equal lines radiating from the centre is the circumference of a circle. The surface of the water, $\beta\epsilon\gamma$, will therefore be spherical.

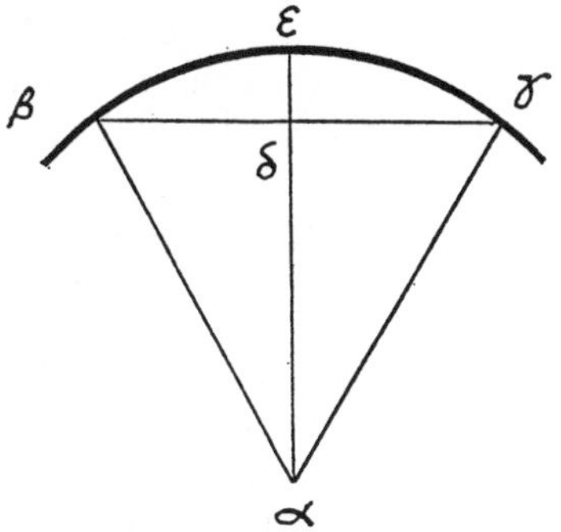

Fig. 3. Aristotle's scheme to prove that the free surface of water is a sphere

Aristotle was a pure theorist. For nearly twenty years of his life, from the age of 17 to 36 years, he was a member of Plato's Academia; he was, therefore, quite naturally permeated by Platonism to a degree which, perhaps, no other great philosopher and scientist has been permeated by the thought of another. This does not mean, of course, that Aristotle agreed with every doctrine of Plato, or studied no new fields. On the contrary, his disagreements with Plato were both violent and notorious, while his epistemology goes far beyond his teacher's horizon.

The birth of fluidstatics

Marcus Tullius Cicero (106–43 BC), Roman orator, statesman and philosopher, wrote in his *Tusculanae Disputationes* (Book V, section 23) that he had found the tomb of Archimedes (287–212 BC) in Syracuse. But what made such a distinguished man endeavour to find the tomb? The answer is really simple says Cicero: I wanted to extract from oblivion the name of a man of humane and noble education, who dedicated his life to research into mathematics and inventions, who contributed more to knowledge than anyone else (by that time, of course).

In 213–212 BC, under General Marcus Claudius Marcellus (268–208 BC), there was a bloody war of conquest by the Romans against Syracuse. Marcellus was a good commander; but Syracuse's resistance was even better and lasted a whole year, because it was organized and led by Archimedes. One day the Romans charged from all sides; the Syracusians thought that this was the end; but then, suddenly, Archimedes activated his military machines: stone throwers, wooden missiles, water-wave-makers. The enemy suffered heavy losses and had to retreat. A week or so later the same tactics were used, then again and again. The Romans, accustomed to quick triumphant victories, had no alternative but to fight for a whole year.

When, at long last, Syracuse fell (212 BC), Marcellus was ordered to find its brilliant defender, to spare his life and to use his inventiveness in the interests of the Roman forces. Archimedes, in the meantime, continued his investigations. One day, while he was regarding a geometrical diagram drawn in sand on the floor of his home (that was the usual way of drawing figures) a soldier entered his study. There was an argument: the soldier wanted to take him to Marcellus, but Archimedes replied that he would go only after he had solved his problem. The soldier began spoiling the diagram: Archimedes demanded that he get off the diagram. The soldier, ignorant of who the old man was, snatched out his sword and killed him.

This was one of the greatest tragedies in the history of science. For, as Leibnitz put it, 'Qui Archimedem et Apollonium intelligit recentiorum summorum virorum inventa parcius mirabitur' – he who mastered Archimedes' and Apollonius' creations, will not be surprised by the discoveries made by the greatest men of our time. Archimedes was, indeed, a towering figure of eternal magnitude. His mathematical investigations led him to a number of very great discoveries and inventions, which revolutionized the pattern of scientific thought. His *Measure of the Circle*, *Quadrature of the Parabola*, *Sphere and Cylinder*, *Conoids and Spheroids*, *Mechanics*, *Hydrostatics*, and other books, articles and inventions were so new, so original, so clever and, above all, so beautifully accurate, that the world of knowledge accepted him at once – in his lifetime – as an outstanding expert, as the founder of Mechanics and Fluid-mechanics.†

Archimedes was, perhaps, the first to examine the internal structure of liquids. He has written very little on the subject, but the little that he

† *Archimedes quae supersunt omnia cum Entocii Ascalonitae commentariis. Ex recensione J. Torelli, Veronensis cum nova versione latine*, Oxford: 1792. *Des unvergleichlichen Archimedis Kunst-Bücher*, Nürnberg: 1670. *Oeuvres d'Archimede, traduites litteralement avec un commentaire par F. Peyrard*, Paris: 1807.

has produced proved to be of paramount importance. First of all, he postulated that, by their very nature, fluids cannot have internal 'empty spaces', that is, they must be continuous. And 'if fluid parts are continuous and uniformly distributed, then that of them which is the least compressed is driven along by that which is more compressed'. Here we have two important concepts of classical Fluidmechanics: (1) pressure applied to any part of a fluid is transmitted to any other part of that fluid, and (2) a fluid flow is caused and maintained by pressure forces.

The famous Archimedes' theorems (or Propositions) follow from here almost automatically; they are:

(1) If a body which is lighter than a fluid is placed in this fluid, a part of the body will remain above the surface (Prop. IV);

(2) If a body which is lighter than a fluid is placed in the fluid, it will be immersed to such an extent that a volume of fluid which is equal to the volume of the body immersed has the same weight as the whole body (Prop. V);

(3) If a body which is lighter than a fluid is totally and forcibly immersed in it, the body will have an upward thrust equal to the difference between its weight and the weight of an equal volume of fluid (Prop. VI); and

(4) If a body is placed in a fluid which is lighter than itself, it will fall to the bottom, because it will be lighter by the amount of the weight of the fluid which has the same volume as the body itself (Prop. VII).

What were the assumptions and postulates upon which Archimedes constructed the beautiful edifice of his Propositions? He assumed, quite rightly, that equal weights of fluids or of a fluid at rest at equal distances from the centre of the earth, are in equilibrium in respect to each other; and he accepted all the consequences of this fundamental postulate. He also proved that when unequal weights are suspended at unequal distances, they, too, may be in equilibrium, but, in this case, the greater weight will have to be at the shorter distance. These and similar assumptions, postulates and theorems were essential for the development of his purely Fluidstatic Propositions.

The continuity-pressure concept of Archimedes, which has been discussed earlier, allowed him to deduce that each part of the fluid is compressed (is under pressure, perhaps, would be better) by the fluid which is above it. Starting from here, he proved a number of Propositions which, in turn, proved that any fluid must have a spherical surface (Aristotle's problem).

If a body whose weight is equal to that of the same volume of a fluid is immersed in that fluid, it will sink until no part of it remains above the surface, but will not descend further.† To prove this Proposition, let a body have the same weight (heaviness) as the fluid. Assuming that this is possible, suppose now that the body is submerged only partly, i.e. a part of it remains above the surface of the fluid at rest. Suppose a plane which passes through the centre of the earth intersects the fluid (Figure 4) and

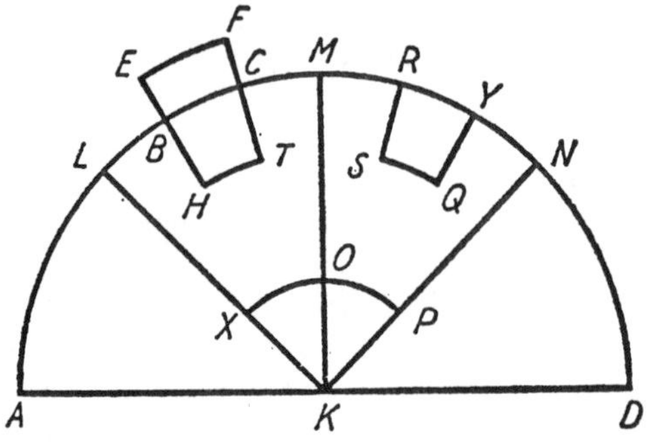

Fig. 4. Archimedes' scheme to prove that if a body whose weight is equal to that of the same volume of water is immersed, it will sink until no part of it remains above the surface, but will not descend further

the body partly immersed in it in such a way that the section of the fluid is *ABCD* and the section of the body is *EHTF*. Let *K* be the centre of the earth, *BHTC* the immersed, and *BEFC* the non-immersed parts of the body. Now draw a pyramid whose base is a parallelogram on the surface of the fluid and whose apex is in the centre of the earth. Let the intersection of the faces of the pyramid by the plane containing the arc *AMD* be *KL* and *KM*. In the fluid, and below *EFTH*, draw another spherical surface *XOP*, the point *K* being its centre, in such a way that *XOP* is the section of the surface by the plane containing the arc *ABCD*. Take another pyramid equal to the first, with which it is contiguous and continuous, and such that the sections of its face are *KM* and *KN*. Suppose that there is in the fluid another body, a 'solidified fluid' *RSQY*, by its size and configuration identical to the submerged part *BHTC*. The portions of the fluid which are contained by the surface *XO* in the first pyramid and the surface *OP* in the second pyramid are equally placed and continuous with each other; but they are not equally compressed, for the portions of fluid contained in *OX* are compressed by the body *EHTF* and also by the fluid contained by the surfaces *LM*, *XO*, and those of the pyramid. But the weight of the fluid contained between *MN* and *OP* is less than the combined heaviness of the fluid between *LM*

† *Des unvergleichlichen Archimedis Kunst-Bücher*. Nürnberg: 1670.

and *XO* and the 'solidified fluid'. For the 'solid' *RSQY* is smaller than the 'solid' *EHTF* – *RSQY* is equal to *BHTC* – and it has been assumed that the immersed body has the same heaviness as the fluid. If, therefore, one takes away the parts which are equal to each other, the remainder will be unbalanced. Consequently, the part of the fluid contained in the surface *OP* will be driven along by the part of the fluid contained in the surface *XO*, therefore there will be no state of rest. Thus, no part of the body immersed can remain above the surface. However, the body will not sink deeper, because it has the same heaviness as the fluid and the equivalently placed parts of the fluid compress it similarly. (*A History of Mechanics*, by Réne Dugas, Routledge and Kegan Paul, 1957.)

The reader will no doubt agree that this and other proofs are somewhat loquacious, repetitive and logical rather than mathematical. But the general stream of Archimedes' arguments was and remains as healthy as Fluidstatics itself. And it could not be otherwise: his conclusions followed almost automatically from his initial continuity-pressure transmission concept. It should, however, be said at once that the final formulation of the law of pressure transmission in fluids was given not by him but by Stevin and Pascal, as will be shown later on in this book.

One of the most celebrated anecdotes concerning Archimedes' Propositions runs as follows. Hiero, the King of Syracuse, had given some gold to a goldsmith to make into a crown. The crown was delivered, made up, and of the proper weight, but it was suspected that the workman had appropriated some of the gold, replacing it by an equal weight of silver. Archimedes was thereupon consulted. Shortly afterwards, when in the public baths, he noticed that his body was pressed upwards by a force which increased the more completely he was immersed in the water. Recognizing the value of this observation, he rushed out, just as he was, and ran home through the streets, shouting ευρηκα! ευρηκα! – I have found it! I have found it! Indeed, he had found it. His experiments showed that water permits an exact computation of solid bodies, because the volume of water displaced weighs precisely the same as the weight lost by the object immersed in the water.

There is, of course, the technical question: how did he do it? And careful reading of various sources suggests the answer: he probably determined the volume of water displaced by the amounts of gold, silver and the whole crown, then calculated the amounts of gold and silver in the crown.

One of the most interesting hydraulic inventions of Archimedes was his Archimedes Screw, a water elevating machine (Figure 5), which could be (and is) used for a number of purposes: to extract water from rivers for irrigation, to pump lubricants in machines, to mix and to supply concrete mixture, etc., etc.

Fig. 5. Archimedes' Screw – a water elevating machine

We thus see that Archimedes occupies a very special position in the history of Fluidmechanics. There are no textbooks of Fluidmechanics which do not begin from or include Archimedes' Law or Archimedes Principle (when a solid body is wholly or partly immersed in a fluid, it is pressed vertically upwards by the fluid with a force equal to the weight of the fluid displaced, the force is known as the buoyancy). For this reason alone, which is not the only one, the notion 'father of Fluidstatics' is fully applicable to Archimedes.

Hero of Alexandria

It is obviously impossible to describe the contributions made by all those who have tried to study fluids. But it is equally impossible to omit any of those who planted the main trees of the subject and became true milestones in its birth and formation. One such man was Hero (Heron) of Alexandria, a Greek scientist and engineer of about the third century, author of numerous works in mathematics, mechanics and physics.†

But why a Greek scientist of Egyptian Alexandria? The answer is as follows. Alexander the Great of Macedonia (356–323 BC) aimed at

† For instance, his *Mechanica*, *Geometria*, *Geodaesia*, *Pneumatica*.

nothing less than the creation of an all-world monarchy or empire. In 332 BC, he founded a new town, Alexandria, on the flat land where the Nile flowed into the sea, which was to be the most magnificent city in the world, the future capital of the monarchy. This designation attracted to Alexandria many outstanding philosophers, scientists, engineers and painters of the time. But after his death, the kingdom was divided among all who could lay hands on any part of it. Egypt fell to the lot of one of his generals, Ptolemy, who chose the still unfinished Alexandria as his capital. Now, this man Ptolemy, a very intelligent but ambitious man, wanted his capital to become a centre of cultural life. To this end, he built an arts museum, library, university and other establishments. Euclid (probably 330–270 BC), was a curator and librarian of the mathematics department of the library; Archimedes studied at Alexandria University; great astronomers like Aristarchus (310–230 BC), Erathosthenes (276–195 BC), Hipparchus (190–120 BC), and others were associated with Alexandria. And so was Hero, yet another mathematician, and hence his name Hero of Alexandria. In a way, and not entirely unsuccessfully, Hero imitated Archimedes. Like Archimedes, geometry was for him mainly a means for the solution of practical problems. It would perhaps be accurate to say that Hero was the first to work out geometrical and trigonometrical methods for determining the heights of trees and mountains and put forward propositions on the determination of the volumes of physical bodies of various configurations.

You ask if Hero was the first engineer? Probably not. It is known, for example, that India and China had highly developed civilizations at least three thousand years before Christ: could this be possible without science and engineering? The Chinese kept records of observations of comets and storms which suggest strongly that they must have had applied scientists and engineers. The written history of China, and to a lesser extent of India, also gives ground to the assumption that hydraulic and melioration engineering must have existed at least in some degree. It is also known that some time around 3000 BC the communities cultivating the alluvial plains of the Nile, and the Tigris and Euphrates were constructing irrigation canals and sails to propel boats. Moreover, the Summerians had drained the marshes along the lower Euphrates near the Persian Gulf and were irrigating land by means of canals long before 3000 BC. Engineering skills enabled Mycenaeans (Mycenae – a famous Greek city from 1400 to 1100 BC) to control flood waters and divert streams. At any rate, the first known civil engineer was Euphalinus of Megara.

Hero's role in the history of fluidmechanical engineering is, however, unique. He gave something like one hundred basic ideas, sketches and drawings, often with detailed descriptions of machines, parts of machines models, measuring instruments, screws, cylinders, pistons, valves, jugs, etc. But, unfortunately for him, he was ahead of his time and many of his inventions were not recognized by the epoch in which he lived.

The more one thinks about Hero's inventions, the clearer becomes the conviction that he must have had a highly logical, inquisitive mind. For example, he must have asked himself: when heated adequately, wood burns, iron melts, stone disintegrates, but what happens to water, to gases, etc.? To get the answer, he probably boiled water in a closed vessel of some kind until it burst. After that, he probably repeated the experiment but with a hole in the vessel and noticed that a whistling stream of steam was shot out through the hole. Now, although I have no material proofs, it is easy to imagine that his vessel was suspended by a chain therefore the hole could not be at its uppermost point. Nor could it be anywhere underneath since it had to contain water; the hole could be made only where it is shown in Figure 6. The steam jet creates a reactive

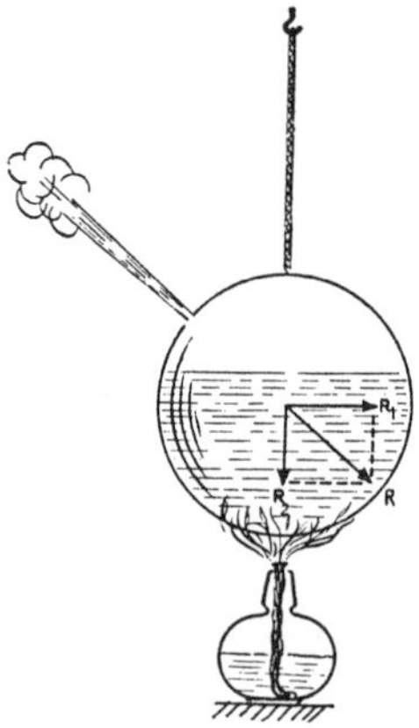

Fig. 6. The discovery of the reactive force by Heron of Alexandria (imaginary scheme)

force R in the opposite direction. Its component R_1 pushes the chain-suspended vessel from left to right, i.e. gives a reactive motion. Having established this revolutionary fact, he could, and did, establish the next revolutionary fact, which became associated with his name. As a good geometrician, he probably asked himself: the line of action of the rocket force R (and its components R_1 and R_2) passes through the centre of the vessel, but what would happen if it did not? He answered his question by fitting a short pipe to the hole as shown in Figure 7. As soon as the

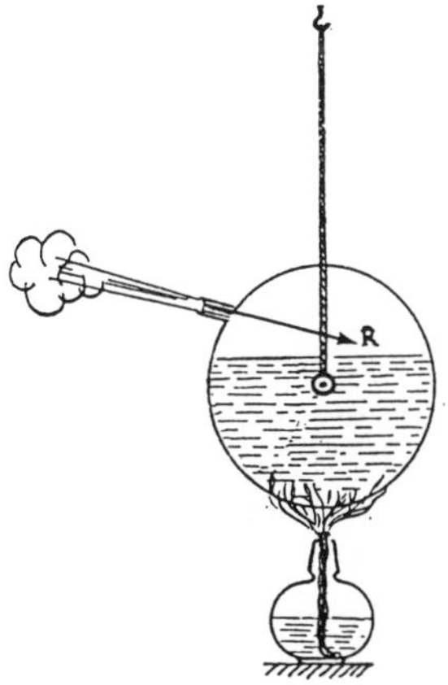

Fig. 7. The discovery of the reactive motion by Heron of Alexandria (imaginary scheme)

water began boiling and the steam jet shooting through the tube, the vessel started rotating in the opposite direction.

The scheme of a suspended vessel, however, had a serious shortcoming: when the hole (tube) moved under the level of the water, it would pour out. To avoid this, the vessel had to be made either to spin horizontally, or to contain steam, and not water. The rest was a mere technicality. To make the machine spin evenly, Hero attached two or three or four exhaust arms to it, as shown in Figure 8. So, something

Fig. 8. Heron's Reactive Motor

very important was born: the future turbine, the future (Bakers) watermill, the future jet engine, the future rocket motor. Simplicity is, indeed, the mother of invention.

Through the Dark Ages to the Renaissance

We thus see that Fluidmechanics has its roots in the very early ages. And although the emergence of its branches had never been smooth, it was gaining momentum. But the worst enemy of man's achievements had always been, and still is, man himself. From the time of the fall of the Roman Empire, the wheel of the history of civilization began spinning slower and slower, until it almost stopped completely, and in some aspects of life even reversed the direction of its rotation.

The barbarian invaders, Goths, Vandals, Lombards and Franks, of Northern Europe, like flocks of crows, hurled themselves upon the fringes of the empire and dragged apart or destroyed almost all that could be classed as civilization. The Barbarian conqueror understood nothing of the Roman craft of government, and so in place of the Roman Law that once had reigned from Tees to Tigris, there grew up chaos, violence, bloodshed and misrule; ignorant of the science of engineering, how could the barbarians keep in repair the Roman walls, roads, bridges, aqueducts? Ignorant of books and writing, how could they understand that the manuscripts being tossed on to the bonfire and dust-heap were storehouses of knowledge? Ignorant of any but the ruder craftsmanships, how could they appreciate the value of the paintings, vases or the beautiful statues that were gleefully tossed out of windows or made targets for skill in archery or spear-throwing? Thus, the clock was set back a thousand years, and the Dark Ages settled down over a whole continent.†

It was the ending of this epoch in man's history which gave birth to the French word *renaissance*, *vozrozhdeniye* in Russian, 'revival' or 're-birth' in English. Needless to say, there was no sudden renaissance; nor was it the product of the efforts of any one man, or even of one nation. No, the Renaissance was the Day that could temporarily be replaced by Night, but could not be prevented from returning again. Yes, the Dark Ages were quite dark. But they were not in unrelieved black. There were many who revolted against the domination of the barbaric order of things and blind authority, thus paving the way for the Renaissance. The struggle was long but it succeeded. The long dark winter broke before the winds that blew upon Italy out of the land of the old Greeks, and the gloom that clouded the spirit of man lifted as the light of the New Learning

† *A Short History of the World.* Elizabeth Underwood, W. & R. Chambers, Ltd, Edinburgh and London: 1959.

spread abroad. Europe began to re-discover the stored knowledge of old Rome and ancient Greece, buried in the dust of Rome's fall.

You have probably read at least something of the writings of Alighieri Dante (1265–1321), the great Italian poet, the creator of the Italian literary language. In his famous *Inferno* (*l'Inferno*), he wrote:†

> Midway this way of life we're bound upon
> I woke to find myself in a dark wood,
> Where the right road was wholly lost and gone.

He was, perhaps, the last great poet of the Dark Ages who lost his right road, but was also looking for new roads. Poetic expressions like 'I once was man', 'false gods were worshipped ignorantly', 'pride fell ruined and ablaze', and so on, showed what he thought of the Dark Ages. But, at the same time, his poems were full of inspiration. His interest in the forces ruling society and nature were deep and immense; the drawings accompanying his writings show this unmistakably. And, to surprise you, I am going to tell you that he was interested also in fluid-mechanics – in matters of importance to future Fluidmechanics, that is.

But first a few words about Greek mythology. There are fantastic stories in it which affected the thinking of very many historic personalities. Somewhere far away to the south, on the shores of the great river Oceanus, according to some myths, were the beautiful Isles of the Blest, where mortals who had led virtuous lives, and had thus found favour in the sight of the gods, were transported without the testing of death, and where they enjoyed an eternity of bliss. These islands had a sun, moon and stars of their own, and were never visited by cold wintry winds.‡

In the myth of the wind, Mercury, or Hermes, was one of the principal personifications. He stole away the cattle of the sun (the clouds). Mercury was the 'lying, tricksome wind god', who invented music, for his music was but 'the melody of winds'. Another personification of the winds was Mars (or Ares), whose name, as I have already pointed out, came from the same root as Maruts, the Indian god. It was first applied 'to the storms which throw heaven and earth into confusion'. Then there was the cloud myth, which said, among other things, that 'the sky itself was a blue sea, and the clouds were ships sailing over it' under the 'terror of the wind'. While Titan stole fire from heaven and bestowed it upon mankind, a clear reference to lightning ('the celestial drill which churns fire out of the clouds').

Thus, obviously, the weather and climate were of great concern to the

† See, for example, his *The Divine Comedy* (translated by Dorothy L. Sayers). The Penguin Classics, 1950.

‡ *The Myths of Greece and Rome*. H. A. Guerber, London: 1910.

Greeks. But they failed to create the science of meteorology. They had created, however, a certain basic model of the world, a kind of Alma Mater, which guided the investigators in a certain direction. Dante was one of them. He accepted the four-elements theory and adamantly advocated that the four elements existed in a spherical shape. The centre of the system, the earth, was entirely covered by an imaginary liquid element; the third element, the air, the atmosphere, clung over the sphere like a soft spherical coat and finally, the fourth element, fire, enclosed the atmospheric sphere. The water blanket of the earth was cool, while the fire was hot, therefore the atmosphere between them was exposed to two diametrically opposed elements, which caused it to be active, rigorous, restless. Rain, hail, storms, winds, etc. were the consequence of this restlessness. This did not mean to Dante, of course, that the atmosphere could not remain motionless. But since the spheres are not isolated from each other by impenetrable spherical walls, motions in one of them effect the state of rest of the others, so that from time to time the spherical structure of the world is distorted.

It would, I think, be right to deduce from this that Dante tried to explain the physical reasons for winds and storms and rains and so on; and, however naïve his theory, he made at least a step forward from the mythological tales. One of Aristotle's books was called *Meteorologia*; Theophrastus of Eresus' (371–286 BC) book was called *On Winds and Weather Signs*; generally the Greeks paid much attention to what we call today meteorology; but they produced no meteorology proper, their theories and philosophies of nature often being dominated by the spirit of myths and mysticism.

Dante was neither a scientist nor a philosopher in the Platonian or Aristotelian sense. But he was a great poet and his poetical inquisitiveness led him through deep valleys. He produced nothing like meteorology, but his thesis that the atmosphere was *not* in a state of permanent tranquility, that there were movements and processes in it (however inaccurate his theory about their origin), made him a figure of interest to future scientists and meteorologists.

We now turn to another post-Dark Age man. Nicholas Krebs Cusa (1401–64), a scientist who lacked the usual scientific logic; a humanist who never had time to finalize his humanistic thoughts; a philosopher whose Crystal Palaces never rose to the intended heights; an advocate of religious reconciliation who failed to build solid bridges of brotherhood: in short, a man who never had time to put his numerous theories and ideas into a coherent and complete system. This very same Cusa suggested the following method for the determination of the humidity of the atmosphere: put on one side of a balance a quantity of wool, and on the other side stones, and make them of equal weights; when the weather is humid the wool will be heavier; but the weight of the stones will remain the same; thus it will be possible to determine the humidity of the air . . .

General remarks about Leonardo da Vinci

Meteorology† is the science of the phenomena of the atmosphere, and Leonardo da Vinci (1425–1519) was the first man to approach its problems scientifically, or nearly so.

The range of Leonardo's interests embraced General Philosophy, Anatomy, Physiology, Natural Philosophy, Optics, Acoustics, Light, Astronomy, Physical Geography, the Atmosphere, the Physics and Mechanics of the flight of birds, the Philosophy and Mechanics of Flying Machines, Mathematics, the Nature of Water, Canal Building, Military Engineering, the Philosophy and Techniques of Painting, Architecture, Music, and so on. When one reads his papers‡ and studies his paintings, drawings, etc., there is no room in one's mind to doubt that Leonardo da Vinci was not only the great figure of the Renaissance epoch, but also the most gifted among the thinkers, engineers, scientists and painters–combined in one person–in the entire history of civilization.

Philosophically and epistemologically, Leonardo's position was hundreds of years ahead of his time. All his thinking revolved, really, around the formula that nature gives man a chance to create technically and artistically; and that the unlimited world of man's creation is his very own. He insisted, perhaps wrongly, that the artist is the true philosopher in that he recreates nature in more quintessential terms, as it were; he extracts from nature more than can immediately be seen by eye, or the mind either. 'No knowledge can be certain, if it is not based upon mathematics or upon some other knowledge which is itself based upon the mathematical sciences', he insisted 'Instrumental; or mechanical science is the noblest and above all others, the most useful.'

This strong reference to mathematics was not accidental, of course. 'Let no man who is not a mathematician read the elements of my work',

† *Meteoros* = that which moves in the atmosphere + *logos* (teaching).

‡ *The Literary Works by Leonardo da Vinci*. Jean Paul Richter, vol. 1, p. 112, Oxford University Press, London: 1939. Also *Leonardo and the age of the eye*, Ritchie Calder, William Heinemann Ltd, London, 1970.

he warned. By which he meant that nobody could be a mechanical scientist, an optician, an astronomer, a fluidmechanics man, a painter or even a poet (Leonardo himself was all these and more in one person) without a certain basic knowledge of mathematics. 'I know', he wrote, 'that many will disagree with me; moreover, many will say that my work is useless, but they will be those who desire nothing but material riches and are absolutely devoid of wisdom which is the food and the only true wine of the mind'. They are those, he repeated on several occasions, who are in love with crude practice without knowledge, therefore resemble the sailor who gets into a ship without a rudder, and who never can be certain where they are going.

No, asserted Leonardo, practical knowledge must always be founded on sound theory. A painter must know the meaning of such notions as 'point', 'line', 'surface', 'volume', etc. He must know that the smallest natural point is larger than all mathematical points because the natural point has continuity, and anything that is continuous is infinitely divisible; but the mathematical point is indivisible because it has no size. In saying that the natural point has continuity, Leonardo means that it has body, mass, and volume.

We have already shown that the concept of continuity emerged in the works of Aristotle, Archimedes and others. But Leonardo da Vinci attempted to give it a broader mathematical and, in due course, physical basis. Imagine, for instance, two adjacent particles in some liquid. They are physical bodies, they have volumes and, consequently, surfaces. How should these surfaces be treated mathematically: as discontinuities? Today the question sounds naïve; but in Leonardo's time it had to be answered quite seriously. And this is how he answered it:

The surface is a limitation of the body. And the limitation of a body is no part of that body: the limitation of one body is that which begins the limitation of another body; therefore that which is not part of any body is a thing of naught; a thing of naught is that which occupies no space. For a surface consists of an infinite number of lines, and a line consists of an infinite number of (mathematical) points, which, being indivisible, occupy no space, and that which occupies no space does not exist . . . Therefore, none of the limitations of two adjacent (fluid) bodies is a part of any of them . . .

It is interesting to add that this concept is repeated by Leonardo time and again, and applied beautifully to the theory of the Linear Perspective of Painting. But to us it means simply this: an imaginary division of any fluid into bodies does not distort its continuity, which is a concept of basic importance to the whole physical and mathematical edifice of Fluidmechanics.

In the same theory of Linear Perspective, Leonardo made a number of references to one of the fluids, the atmosphere. 'That the atmosphere attracts to itself all the images of the objects that exist in it, and not their forms but merely their nature', he says, 'may be clearly seen in the sun, because the whole atmosphere becomes completely shot through with light and heat (of the sun). It can clearly be shown that all bodies are, by their images, all in all the surrounding atmosphere. Every body in light and shade fills the surrounding air with infinite images of itself . . . The body of the atmosphere is full of infinite pyramids composed of radiant straight lines, which are caused by the boundaries of the bodies.' And then one of the fundamental natural facts upon which a whole branch of modern Fluidmechanics is built: 'Just as a stone flung into the water becomes the centre and cause of many circles', says Leonardo da Vinci, 'and as sound diffuses itself in circles in the air, so any object, placed in the luminous atmosphere, diffuses itself in circles and fills the surrounding air with infinite images of itself.'

And so on, and so forth. The great painter and engineer was also a great geometrician, physicist and optician, who clarified a large number of riddles associated with water and air. The beautiful geometrical pictorial illustrations accompanying his writings† convincingly show that all his plants had deep roots.

But we have not yet touched upon his really major works. It is difficult to imagine a more brilliant and more immortal scientific work than Leonardo's *On the properties and on the movements of the human figure*. In view of the fact that there were at that time no anatomy or other associated sciences, it must be confessed that he opened up not the horizon but the whole sky of knowledge. When you analyse his writings and look at his accompanying illustrations, your intellectual pulse races, your amazed mind is carried into an era of wonder, your scientific imagination finds itself on wings. Could he be real? Yes, he was real, and here he is, before your eyes. Goethe once said that good architecture is frozen music, well, Leonardo's works are the unsurpassed spring of living music!

And what about his *Botany for painters and elements of landscape painting*? Has anyone ever produced a work of such completeness? Did not the subject play merrily in his mind and through his amazing writings? Again, you read and your mind recalls Dimitri Merezhkovsky's words about Michelangelo, that his strength was like the beauty and completeness of his genius, like a tempestuous wind, rending mountains, and breaking rock in pieces before the Lord; and that he, Leonardo, was

† See also, for example, *Leonardo da Vinci*. Tagebücher und Aufzeichnungen, Leipzig: 1953.

even stronger than Michelangelo, as the calm is stronger than the tempest.†

Philosophers, however, often denied him, the 'man without letters', a place in their midst. Yes, of course, there are sometimes philosophical incoherences in Leonardo's theories. But the careful reader of his immortal works cannot fail to compare him with *force*, which he defined as 'an incorporeal agency, an invisible power, which drives away whatever opposes its action, i.e. that which conquers and slays the cause of opposition'. That is what Leonardo was: powerful, indefatigable, full of energy, brilliant in imagination, unsurpassed in techniques, inexhaustable as a mighty river with its sparkling source and many tributaries.

Leonardo da Vinci's original works on fluids

Leonardo da Vinci was also a great civil engineer and town planner. He studied, analysed, planned and designed almost everything and anything: roads, streets, canals, bridges, houses, palaces, shopping centres, arcades, water basins, swimming pools, vessels to accumulate rain water, etc. etc. One house must turn its back to another; the roads and streets must have definite widths, high streets must be free of vehicles so as to serve gentlemen and constructed so as to provide the shortest possible communications; provisions, such as wood, water, wine, were to be carried in through doors; privies, stables and other fetid matter must be emptied away underground; cities must be built near the sea or large rivers in order that the dirt (of the cities) may be carried off by the water; a town must be so planned that the inundation of the rivers does not flow into the cellars, does not cover the streets, and does not cause an elevation of the water level of the canals. A typical plan of such a town is shown in Figure 9.

His works on astronomy, mechanical engineering, civil engineering, optics, fluids, etc., give you the feeling that mathematics was sitting in a nearby chair; that beneath every basic conception lies applied mathematics; and that his drawings and schemes are mathematical equations of a special sort. 'Mechanics is the paradise of the mathematical science, because in it we come to the fruit of mathematics', he wrote;‡ 'there is

† *The Romance of Leonardo da Vinci.* Dimitri Merezhkovsky, the Modern Library, New York: 1928.

‡ *Leonardo da Vinci, Leisure Arts*, vol. 1. Instituto Geografico De Agostini, Novara, Italy: 1964.

no certainty as to where mathematical sciences cannot be applied, or where mathematics is not connected with them.'

This explains, at least partly, why his engineering designs were almost always so accurate from the mathematical point of view; and his velocity-area law may serve as an example.

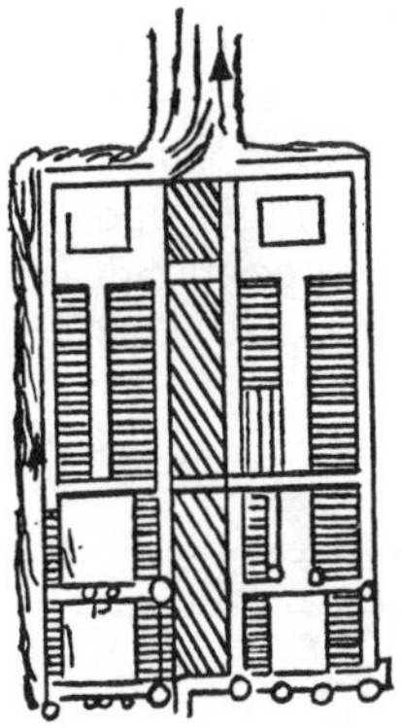

Fig. 9. Leonardo da Vinci's town plan

Leonardo da Vinci was an excellent observer, and his motherland, Italy, was rich with natural phenomena and processes to be observed, such as water-falls of various kinds: a field of knowledge where one discovery led to another. It is obvious from his notes that he planned to write separate books on seas, rivers, laws of motion of water, and methods of measuring the properties of fluids more generally. Here is a typical plan written by him:† 'Divisions of the book: of water itself, of the sea, of subterranean streams, of rivers, of the nature of the depths, of obstacles, of gravels, of the surface of the water, of the things moved therein, of the preparing of rivers, of conduits, of canals, of machines turned by water, of raising water, of matter worn away by water.'

But even Leonardo's life was nearer every day to its end, and so this book, and many other books, never saw the sun. Nevertheless, he left us with more than plans and intentions. His town planning works embraced almost all the basic ideas of water supply. Italy had hot summers, he wrote, during which some rivers dried up completely and some others partly; and this was bad for towns. But what were the remedies? In the first place, towns and villages should be built near those rivers whose water levels did not rise or fall too much. Secondly, man had to learn to control his rivers.

† *The Literary Works of Leonardo da Vinci.* J. P. Richter, London: 1939. Also, *Tagebücher und Aufzeichnungen von Leonardo da Vinci*, Leipzig: 1953.

Why, for example, does the river Arno ruin the walls of its banks?, Leonardo asked. Because its tributaries transport vast masses of soil, sand and stones into it and deposit them on the riverbed, especially on the opposite side, thereby affecting the pattern of the flow, and distorting it. But even if there were no such deposits, the flow energies of the tributaries would disturb 'the natural behavior of the mother river', and 'a disturbed river is merciless'. For instance, the river Mensola not only spoils the bed of the mother river (Arno), but also forms an eddy at its junction. This eddy, being a whirlpool, a mass of fluid in which the flow is circulatory, 'works itself into the opposite side of the bank and spoils it'. On the other hand, if the opposite side is not high enough, an outflow branch may develop (Figure 10). This and other 'violences' can be

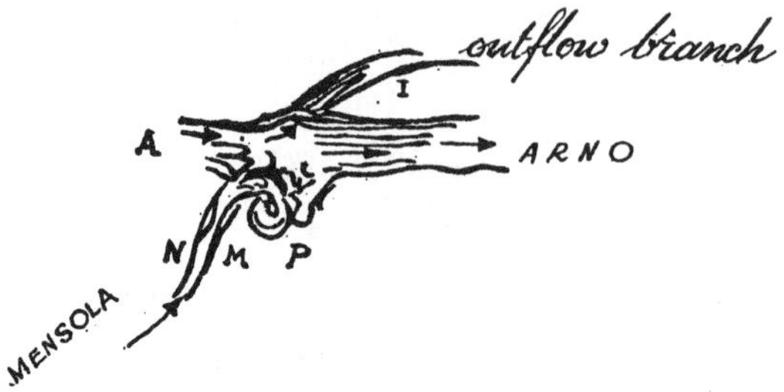

Fig. 10. The river Mensola not only spoils the bed of the mother river Arno, but also forms an eddy and may form an outflow branch

avoided by means of flood gates capable of forcing rivers to join each other in a 'coaxed way'. Beyond the first flood-gate (says Leonardo), at the same distance downstream, there should be another gate, then a third, then a fourth, then a fifth, so that the river may discharge itself into a canal (built for it) coaxially with the mother river.

Why so many flood-gates? the reader may ask. Because, says Leonardo,† if you have only one gate (Figure 11), the distance *AB* is too

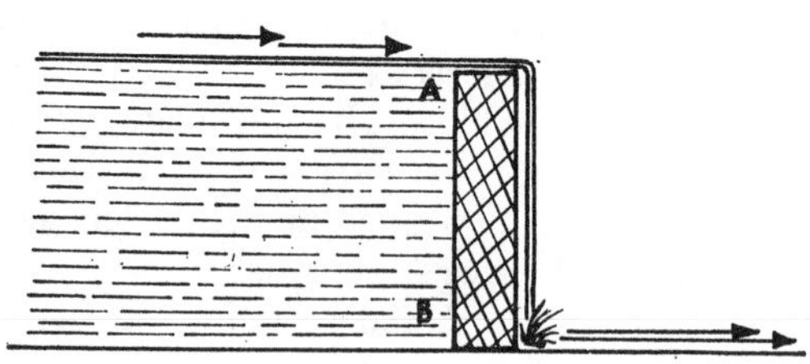

Fig. 11. Falling water hits the lower bed and destroys it

† *The Literary Works of Leonardo da Vinci*. J. P. Richter, London: 1939. Also, *Tagebücher und Aufzeichnungen von Leonardo da Vinci*, Leipzig: 1953.

high; and this is undesirable, because 'the falling water will hit the lower bed very violently' and, consequently, will destroy it slowly. In order to avoid this, the distance *AB* should be divided into several small distances (Figure 12).

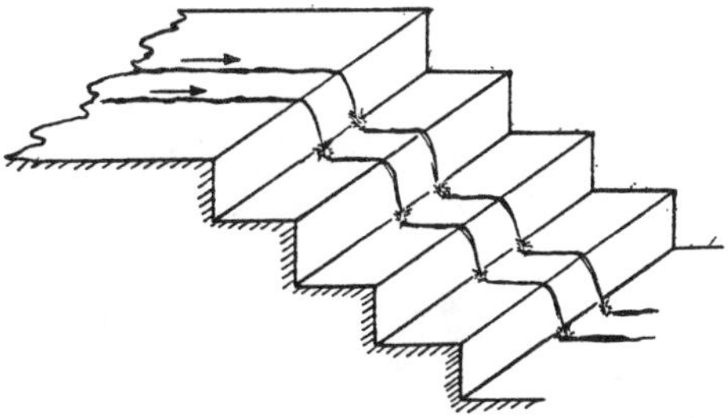

Fig. 12. How to reduce the destructive power of falling water

He then discusses the fate of the river Adda. The water of this river 'is greatly diminished by its distribution over many districts for the irrigation of the fields', therefore people living along its banks experience shortage of water. How is this remedied? he asks, and gives the answer: part of the water of the river 'is drunk up by the earth', or lost; but if we build little channels (under the surface of the earth), the lost fluid will be re-accumulated and become useful to people.

Water in the earth and blood in a living body flow similarly; the veins which discharge blood, are not discerned by their smallness. The waters of the earth return with constant motion from the lowest depths of the sea to the utmost heights of the mountains, not obeying the nature of heavier bodies; and in this they resemble the blood of animated beings which always moves from the sea of the heart and flows towards the top of the head; and here it may burst a vein, as may be seen when a vein bursts in the nose; all the blood rises from below to the level of the burst vein. When the water rushes out from the burst vein in the earth, it obeys the law of other bodies that are heavier than the air, since it always seeks low places.†

What were the laws governing the flow of water and blood? All inland seas and lakes are due to rivers, he says. One of the more important laws of behaviour of seas and lakes is that they, and also oceans, 'display wave motion'; and a wave always breaks on its bed (Figure 13) near the shore. As to rivers, he says, there exists one general law for all cases: *where the flow carries a large quantity of water, the speed of flow is greater and vice versa.* Somewhere else, he formulates the same idea differently:

† *Literary Work of Leonardo da Vinci*, vol. II. Oxford University Press, London: 1939.

Fig. 13. A wave breaks on its bed near the shore

where the river becomes shallower, the water flows faster. This idea can be traced in all his drawings of water, blood, and air flows.

Take, for example, his famous water-mill (Figure 14). It contains

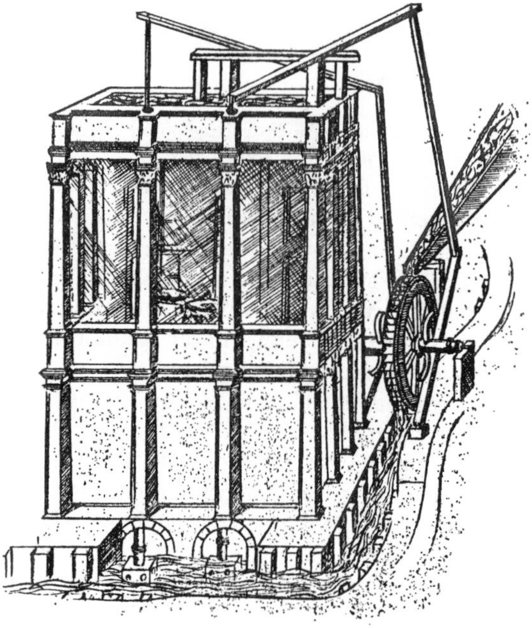

Fig. 14. Leonardo da Vinci's watermill

elements of simple automatic control of the speed of the wheel, has a braking mechanism and (*ad hoc* still more important) the configuration of the water-supply channel suggests that Leonardo exploited the same interdependence of the cross-sectional area of a flow and its velocity, the so-called *velocity-area law*, namely, that the velocity V of a fluid flow is greater where its cross-sectional area A is smaller (and vice versa), the product of the two being constant:

$$\boxed{VA = \text{const}}$$

or $V_1A_1 = V_2A_2 =$ const, or $V_1:V_2 = A_2:A_1$. In other words, if you narrow the cross-section of a flow, you thereby increase the flow velocity at that cross-section. And an increase in V means an increase in the kinetic energy of one and the same mass, i.e. an increase in its working power. We shall be returning to these questions more than once. In the meantime, we can confidently say that the establishment of this law alone puts Leonardo da Vinci among the great fathers of Fluidmechanics.

I have used the words 'working power' above deliberately. Power is the time-rate of doing work, $P = dW/dt \equiv W/t$, where P is power, W is work, d is differential, and t is time; and work is defined as the product of the displacement (say, in metres) by the force (say, in kilograms) in the direction of that displacement.

Hence the question which has been encountered by every scientist and philosopher throughout the entire history of civilization: what is force? And it was here that Leonardo da Vinci revealed his inability to give a good answer.

These are the definitions he gave:† Force is a spiritual power, an invisible energy, imparted by violence from without to all bodies out of their natural balance; an invisible energy, which is created and communicated, through violence from without, by animated bodies to inanimate bodies, giving to these the similarity of life, and this life works in a marvellous way; violence which dies through liberty; that which drives away in its fury whatever stands in its way to its ruin; that which transmutes and compels all bodies to a change of form and place; which is always opposing forces of nature; which is but a desire to fly. Weight does not change of its own accord, while force is always a fugitive; weight has a body, force has none; weight is material, force is spiritual; if one is eternal, the other is mortal.

A chain of semi-mystical alleluias, you may say. But he also says, somewhere else, that, whatever its origin, force keeps the world together. Or, to put it differently, the final results, physical laws, engineering erections, produced by Leonardo suggest that his mysticism did not divert him from the road of science and technology. Besides, some of the above definitions are perfectly scientific.

† *Literary work*, vol. II. Oxford University Press, London: 1939.

Leonardo's fluidmechanics

Let us turn now to Leonardo's more specific contributions to Fluidmechanics. The word Fluidmechanics is used here in the broad sense that has already been outlined. Like his predecessors and contemporaries, he believed that the universe consisted of Earth, Water, Air and Fire. The centre of the watersphere was for him the true centre of the Earth. And in this sphere the mass (today we would say 'the mass density') varies conspicuously from the surface to the centre, but is continuous. Water penetrates all the spaces between rocks, stones, sand grains, therefore no vacuum exists between Earth and Water. Similarly, no vacuum is possible between the atmosphere and fire. So, he says, 'anyone who says a vacuum is generated speaks foolishly'.

Our reader will have a chance to see, later on in this book, that, indeed, the most fundamental law of the universe is its universal matter-energy continuity, that is, there are in the beginningless and endless universe no matter-energy discontinuities, i.e. absolutely empty regions. Leonardo used the old-fashioned word vacuum, but, in fact, he was in line with the modern universal continuity concept.

One of the practical questions he analysed was this: man and animals can live in one of the four elements, in the atmosphere; but why can they not live in water? And he worked out a practical solution. 'I do not wish to publish or divulge it, however', he wrote,† 'because of the devil nature of men, who would use my method as a means of destruction (of each other) at the bottom of the sea . . . I can only say that the end of the breathing system can be attached either to a rock or kept above the surface in some other way.'

He then discusses navigation problems of various kinds. The ancients, he says, used mill-wheels to propel their boats. These wheels touched the surface of the sea by their tips. But if the water moves together with the ship, say in a river, at the same velocity, the wheels must remain motionless; if the water flows faster than the ship, then the work of the wheel becomes too complicated. Why, then, not use another method of propulsion: a board attached to the ship in such a way as to expose it to the pressure of wind? If the board is exposed obliquely, according to the strength of the wind, then the ship can move at a desired speed . . .

The sail was not, of course, unknown before Leonardo da Vinci. But

† *Tagebücher und Aufzeichnungen von Leonardo da Vinci*, Leipzig: 1953.

now it emerged as an engineering design, as a large cloth, usually canvas, spread to the wind to propel a vessel through water. This, not surprisingly, led him to the solution of yet another problem: what should the external configuration of the boat itself be? He gave his answer by sketches and drawings of boats of various types and sizes (Figure 15).

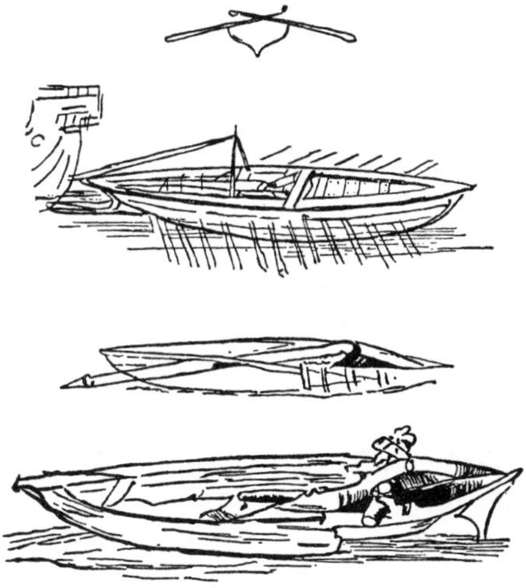

Fig. 15. Leonardo's sketches of boats

Whether one uses propeller-wheels or propeller-boards (i.e. sails), it is important that the ship moves through water with the least possible resistance. Therefore, he says, man should learn from fishes, since they live in water and are shaped to move in it easily. 'It is wonderful to watch how dolphins jump out of water and make leaps without meeting resistance'. The same, he continues, is true of fish and other animals capable of prolonged swimming.

This is what he means (Figure 16): when a body moving in water has

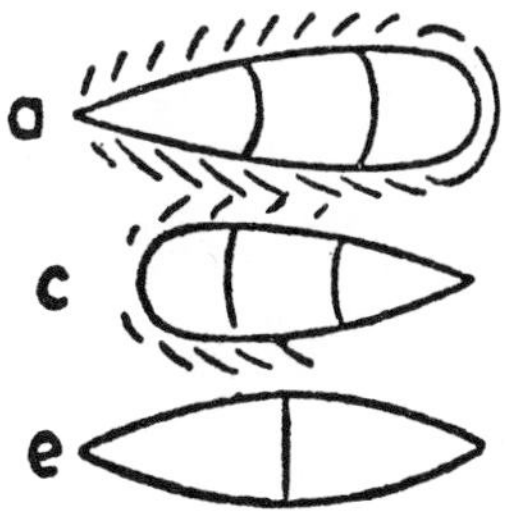

Fig. 16. Leonardo's schemes for the theory of bodies of least fluidmechanic resistance

the configuration (c) and its blunt side is forward, fluid particles move smoothly only over the forward part of its surface; at a certain point, they cease to be 'undisturbed' and reach 'violence': in modern terms, they separate from the surface and form a vortex trail.

If the same body is moving its pointed side forward as in configuration (a), the smooth flow is longer; but, still, there is 'violence', therefore it, too, meets resistance, although smaller than in the first case.

If we now have a body like configuration (e), water particles will remain in smooth motion throughout the whole length, therefore water resistance will be still less than in the case of (a). Finally, the configuration of a fish provides both minimum resistance to motion and manoeuvrability.

The manoeuvrability of a ship, he thought, was very similar to that of a fish. 'Inasmuch as all beginnings of things are often the cause of great results, so we may see a small almost imperceptible movement of the rudder to have the power to turn a ship of great size and loaded with a very heavy cargo, and amid such a weight of water which presses on it from all directions and against the impetuous winds acting upon its mighty sails. Therefore, we may be certain in the case of those birds which can support themselves above the course of the winds without beating their wings that the slight movements of the wing or tail which will serve them to enter either below or above the wind will suffice to prevent the fall of the same birds.'

Generally, according to Leonardo, there is a great similarity between movements of bodies in water and the air. Swimming in water teaches men what birds do in the air. But one must learn first in what way man ought to learn to swim, how to rest on the water, how to protect himself against whirlpools or eddies which tend to drag him down, how a man dragged to the bottom must seek the motions which will throw him up from the depths, and why he cannot stay under water unless he can hold his breath.

Before flying, you must swim, and while learning how to swim, have a safety coat or ring made of leather, advises Leonardo (Figure 17). This

Fig. 17. Leonardo's swimming safety ring

coat should have the part over the breast with two layers, a finger's breadth apart; and in the same way it must be double from the waist to the knee; and the leather must be airtight. When you have to leap into the sea, blow out the skirt of your coat through the double layers of the breast; and jump into the sea and allow yourself to be carried by the waves; keep in your mouth the air tube which leads down into the coat.

As to flying in the air, man should first study the behaviour of the birds. He will then notice several important facts. First of all, the 'flying machine', whatever it may be, should allow man to be free from the waist upwards, so that he is able to balance himself as he does in a boat. Secondly, the centre of gravity of himself and of the machine must counterbalance each other, and be shifted as necessity may demand. Thirdly, the wing, or machine, must not be too heavy, otherwise the air will not be able to support it. Before building wings, man must study the anatomy of the wings of birds, together with the muscles of the chest which move the wings.

Man is, however, far too heavy to fly like a bird, Leonardo warned. His flying machine must imitate nothing other than the bat (Figure 18),

Fig. 18. Leonardo's bat

because the membranes strengthen the whole framework, and provide a larger lifting force. If, on the other hand, you imitate the wings of feathered birds (Figure 19), you will have structure that is much stronger,

Fig. 19. Leonardo's bird

because the feathers are separate, less air passes through them and they do not reinforce each other. But the bat is aided by the membrane that connects the whole and is not penetrated by the air. In short, dissect the bat, and keep to this study, and arrange your flying machine on this model.

What other machines could be used by man for the purpose?, asks Leonardo, and replies: Not many, because the air is 'much thinner' than water, therefore its sustaining force is restricted to 'wing-machines' and 'air-screws'. The latter can be imagined either as a whole Archimedes water-lifting screw, or as a portion of it (Figure 20). If the portion is 'screwed into the air' at a speed high enough, it will create an upward force capable of lifting man into the air. Here we recognize the idea of the modern helicopter.

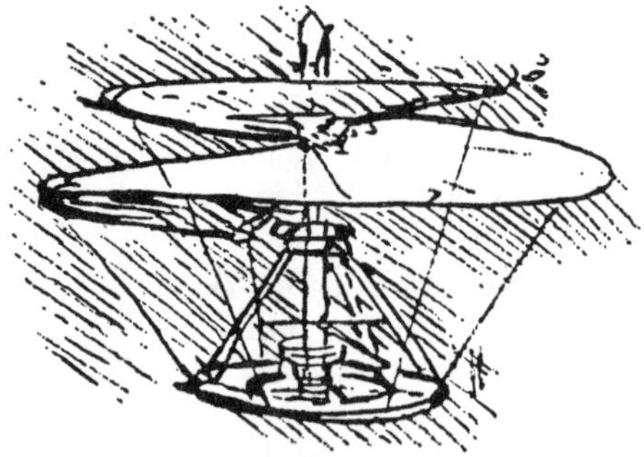

Fig. 20. Leonardo's air-screw, the ancestor of the modern helicopter

See how the beating of its wings against the air supports a heavy eagle in the highly rarefied air; observe also how the air in motion fills the swelling sails and drives a heavily-laden ship. It is clear from these instances, he declares, that a man with wings large enough, and duly attached to his body, might learn to overcome the gravitational pull and the resistance of the air, and conquer the atmosphere.

But suppose the wings break, or the machine turns on its edge and loses controllability: how could the man be saved? asks Leonardo. And he provides four answers, which are today the *A*, the *B*, the *C* and the *D* of flight. In the first place, if the machine turns on its edge at a high altitude, it will have enough time to regain a normal flight condition, provided that its structure is sufficiently strong and its air resistance high. Secondly, all who fly should wear special bags strung together like a rosary (Figure 21) and fixed on their backs. In this way, a man falling

Fig. 21. Leonardo's safety belt made up of airtight bags

from a height may avoid hurting himself. Thirdly, accidents of the kind we are discussing may be caused by winds and changes in the atmosphere, therefore 'we need a clock to show the hours, minutes and seconds to measure what distance per hour one travels with the course of the wind, and to know the quality and density of the air, and when it has to rain'. Fourthly, a *body moving in the motionless air experiences as much air resistance as is experienced by the same body in a motionless state but exposed to an air motion with the same speed* (Leonardo da Vinci's Principle); if, therefore, we attach to a falling man a horizontal sail, his descent will be resisted by the air; still safer, if the sail is made in the shape shown in Figure 22. The explanatory note accompanying the

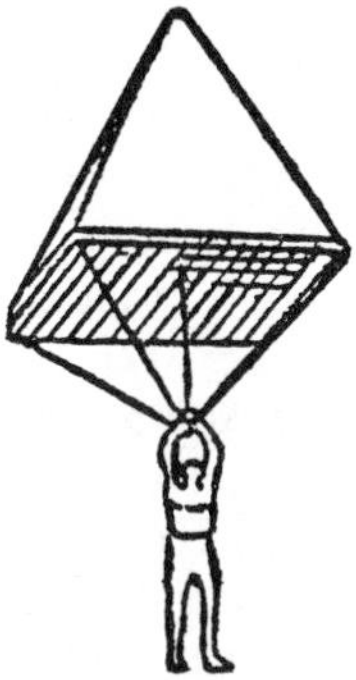

Fig. 22. Leonardo's parachute

drawing says: Wenn ein Mensch ein Zeltdach aus abgedichteter Leinwand, das 12 Ellen breit und 12 hoch sein soll, über sich hat, so wird er sich aus jeder noch so grossen Höhe herabstatürzen können, ohne Schaden zunehmen† (If a man has a tent of linen without apertures, twelve ells across and twelve in depth, he can throw himself down from any great height without injury). And so the idea of the modern parachute was born.

Leonardo's work on the mechanics of flight embraces a wide range of problems. His drawings of birds in various flight situations, and his explanations of them, are so beautifully accurate that they served as a sound skeleton for the mechanics of flight of the aircraft itself.

The air, or the atmosphere, meant to him a continuous and uniform matter composed of molecules. Its brightness is due, he says, to the water vapour which 'has dissolved itself in it into imperceptible molecules'. These, being lighted by the sun from the opposite side, create the brightness which is visible in the air; and 'the azure which is seen in it is caused by the darkness that is hidden beyond the air'.

Now, vapour is water, and water is heavier than air: how, then, does it penetrate the atmosphere? Leonardo gives this explanation:‡ The

† *Tagebücher und Aufzeichnungen von Leonardo da Vinci*, Leipzig: 1953.

‡ *Leonardo da Vinci*, vol. II. Instituto Geografico de Agostini, Novara, Italy: 1964. *The Literary Works of Leonardo da Vinci*, London: 1939.

'return eddies' of wind at the mouth of certain valleys strike upon the waters and scoop them out in a great hollow, whirl the water into the air in the form of a column, and of the colour of a cloud. 'I saw this thing happen on a sandbank in the Arno', he writes. 'The sand was hollowed out to a greater depth than the height of a man; and the gravel whirled around and flung about appeared in the air in the form of a great bell-tower. . . .'

Obviously, Leonardo is referring here to the so-called tornado, a violently rotating column of air, nearly always observable as an inverted truncated cone. It is, perhaps, the most destructive of all atmospheric phenomena. Its vortex, commonly several hundreds of yards in diameter, whirls usually cyclonically, with wind speeds from 100 to 300 m.p.h. Very low pressure prevails inside a tornado because of its great vorticity, and it sucks up not only sand and gravel, but even whole houses. Smaller tornados, or whirlwinds, are usually not more than a few hundred feet high and several tens of feet in diameter. They appear normally in a shallow layer of very unstable air lying immediately above the ground.

Strong winds were not just harmless natural phenomena; no, they often cause great damage to mankind. But could they be prevented, made useful to man? Yes, of course, said Leonardo. For example, if we watch, or measure, the motion of the shadow of a cloud over the land or water surface, we know the direction and the speed of the wind, because it is the latter which moves the cloud. It was impossible, however, to run after the shadow across rivers, seas, rocks and forests, therefore a less troublesome method was needed, and Leonardo sketched it: the obvious predecessor to the modern wind vane (Figure 23). The writing attached

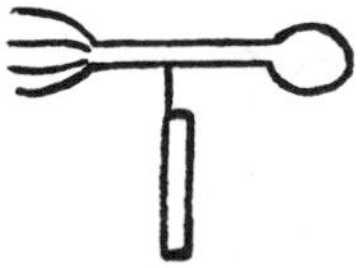

Fig. 23. Leonardo's wind vane

to the drawing reads: *Modi di pesare l'arie eddi sapere quando s'a arrompere il tepo* (mode of weighing the air and of knowing when the weather will change).

Simon Stevin (1548-1620)

One of the great figures of post-Leonardo times was Simon Stevin, often referred to as Stevinus, who was born in Bruges in 1548 and died at the Hague in 1620. He was a military and civil engineer under Maurice of Nassau, the man who introduced decimal fractions into common use. Like Leonardo, he was a man of many talents. He invented a carriage propelled by sails for twenty-eight passengers and published several books, including one called *Statics and Hydrostatics* (Leyden, 1586), in which he attempted, not unsuccessfully, to revise some of the fundamental concepts of his predecessors, and to give them a good mathematical face-lift. In his *Optics*, Leonardo da Vinci developed the triangle concept to the degree of complete perfection. But it was Stevinus who, for the first time (so far as is known), introduced the concept of the triangle of forces (in effect, the concept of the triangle of vectors) or, what is the same, of the parallelogram of forces.

This was, roughly, the theoretical philosophy behind his discovery. Generally speaking, a material body is under the action of various forces acting in various directions. For example, a falling body is under the action of its own weight, air resistance, and the force of wind; the first of these acts vertically downwards; the last acts in the direction of the wind, thus preventing the body from falling vertically; and the force of air resistance always acts in the direction opposite to the direction in which the body moves. The question was, then, how to deal with such a system of forces.

Imagine, said Stevinus, that we have a chain *ABCD* around a fixed edge (Figure 24). With its hanging part *D*, it will be at rest. But if we

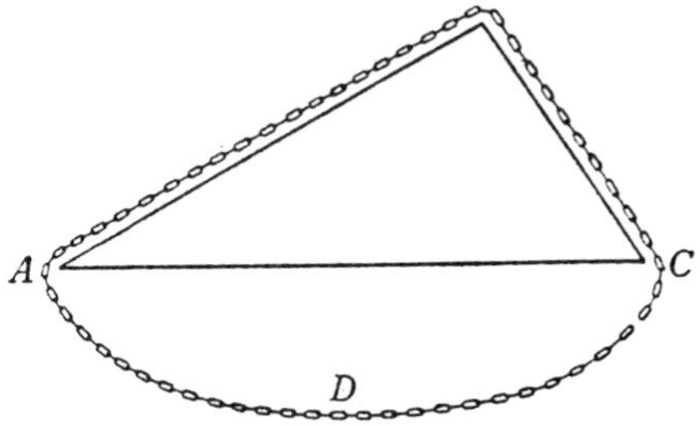

Fig. 24. Stevin's chain

cut away the part of *D* at *A* and *C*, the remaining parts *BA* and *BC* will continue their equilibrium state. Which means that any two bodies on the faces *BA* and *BC* connected by a string would also be in equilibrium, if their weights were in the ratio of the lengths of the faces. From this, simple mathematics led him to the rule of the parallelogram of forces, with the following proof.

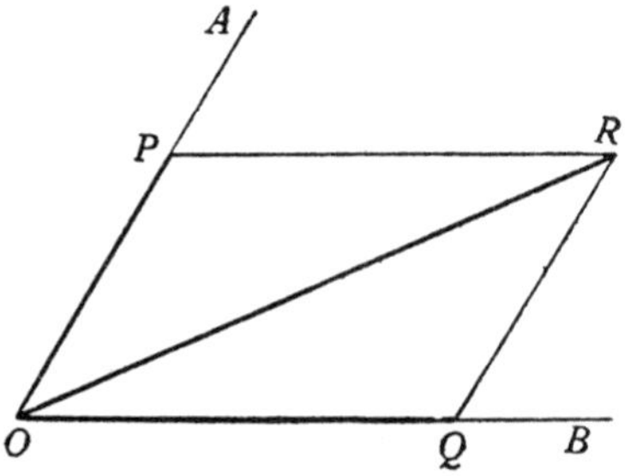

Fig. 25. Stevin's parallelogram of forces

Let two forces of different directions be applied to one point O (Figure 25). Let the lengths $\overline{PO}$ and $\overline{OQ}$ be proportional to the magnitudes of the forces. Then the diagonal $\overline{OR}$ of the parallelogram built on the two forces represents the resultant force, whose effect upon the body is exactly the same as of the forces $\overline{OP}$ and $\overline{OQ}$ acting separately but simultaneously.

I must, however, point out that the idea of resolution of forces in this manner was not unknown to Leonardo da Vinci. He asked himself how the weight of a heavy body supported by two strings was divided between them. And he thought that the weight of the body suspended at *b* was divided between the strings *bd* and *ba* as the ratio of the lengths *ea* and *de* (Figure 26). This was an incorrect inference, but the fact is that Leonardo, too, gave the problem some attention.

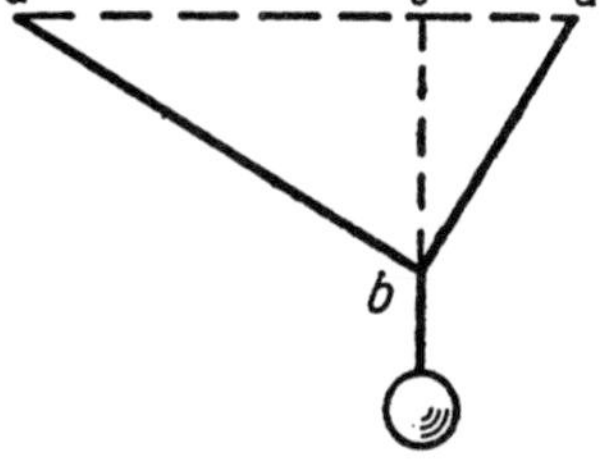

Fig. 26. Leonardo's scheme of resolution of forces

But to return to Stevinus, he made a good attempt, again not an unsuccessful one, to give Archimedes' *Propositions* mathematical and experimental proofs. And while doing this, he enriched Archimedes' Fluidstatics with further examples and illustrations. One of his more important, certainly original, contributions was his 'principle of solidification'. According to this principle, a solid body of any configuration, but of the same mass density as the fluid, can remain in that fluid in complete equilibrium, whatever its position may be, and without disturbing the pressure in the rest of the fluid.

To understand and appreciate the importance of this principle, let us bear in mind that, up to the time under discussion, there were still no mathematical methods for the determination of pressure distribution over bodies immersed in fluids. Stevinus was, we think, the first man to attempt such a method, on the basis of his 'solidification principle'. It was he who established that *pressure is independent of the configuration of the body, and depends only on the weight of the column of liquid above it.*†

The other original contribution by him was the establishment of the so-called 'hydrostatic paradox': the pressure experienced by the bottom of a vessel containing a fluid depends only on the (horizontal) area of the bottom and the depth below the surface of the fluid, but does not depend on the shape of the vessel. This is how he proved it (Figure 27): one and

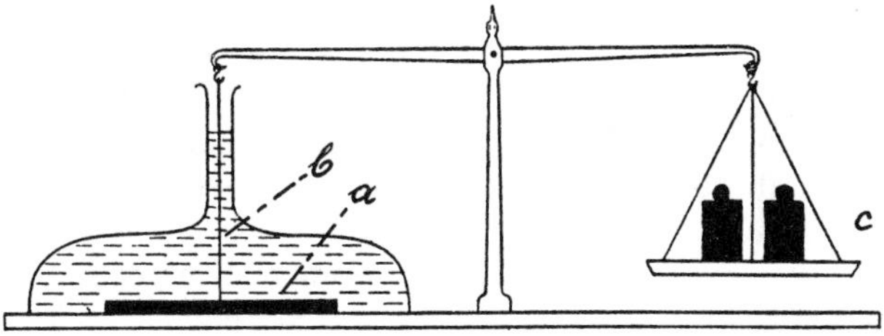

Fig. 27. Stevin's hydrostatic balance

the same light plate (a) was placed under vessels of different shapes, all filled with one and the same water. In each case, the pressure was measured through string (b) by weights (c), and proved to be the same.

Finally, Stevinus described how he and his collaborator Gretius had experimented on the fall of bodies under gravity, and found that a lightweight and heavyweight body dropped from the same height took the same time to reach the ground. This was contrary to Aristotle's theory that bodies of different weights would not reach the ground at the same time.

† *Histoire des sciences mathématiques et physiques chez les Belges.* L. A. J. Quetelet, Brussels: 1866.

Stevinus exploited Leonardo's flood-gate idea to the full and developed, on its basis, a whole system of defence lines, and supervised the building of canals, water reservoirs, etc.

Galileo Galilei (1565-1642)

Archimedes founded Fluidstatics; Leonardo da Vinci and Stevinus gave the first impulse to the renewed study of Fluidstatics; but Galileo Galilei laid the foundations of general Dynamics, without which there would be neither Fluidmechanics nor Mechanics generally. He was unhappy with some of the basic concepts of Aristotle's Mechanics. For example, like Stevinus, he refused to believe that bodies of different weights fall differently, and to prove his point, carried out a series of experiments by dropping various bodies from the same height, probably from the top of the leaning Tower of Pisa, and found that a cannon ball and a musket ball took the same time to reach the ground, thereby disproving Aristotle's doctrine.†

As early as 1604, Galileo arrived at the conclusion that the distances covered by falling bodies are proportional to the square of the times of fall, $s = kt^2$, where k is a constant coefficient; which is one of the classical laws of Mechanics. It followed from here that the distances travelled in equal times were related to each other as the consecutive odd numbers starting from one.

The above law proved to be of fundamental importance both in Mechanics and Fluidmechanics. But his role in Fluidmechanics goes far beyond that. He introduced the concepts 'inertia' and 'momentum' in no uncertain terms, improved the ancient water-clock, improved Stevinus's parallelogram of forces, became known as 'the father of the theory of projectiles', etc., etc. The manner in which he developed his projectile theory is extremely interesting. Imagine that you are throwing a ball along a smooth horizontal plane, he says. Then, assuming that it moves without any obstacle, immediately after reaching the end, the ball will start moving also downwards, under the action of its own weight, i.e. the motion will have a forward and downward component, and the trajectory of the ball will be a parabola (Figure 28). He called this kind of motion projection (hence the name projectile).

† *The growth of physical science.* Sir James Jeans, Cambridge: 1947.

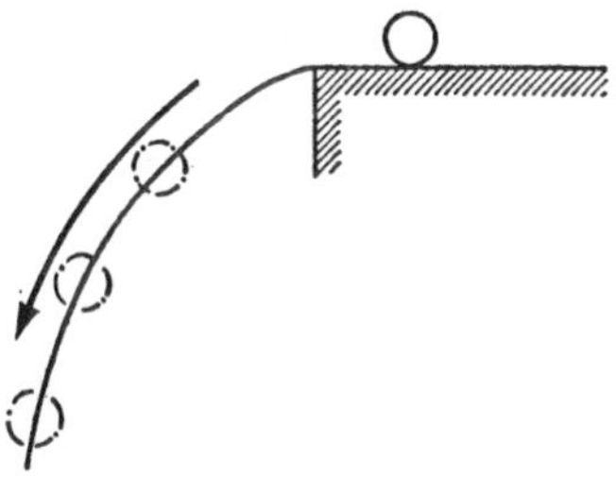

Fig. 28. Galileo's projectile parabola

The proposition was true for fluid particles also. He studied Fluid-mechanics, however, in a broader sense.† In a work called *Discorso intorno alle cose che stanno in su l'acqua o che in quella si muovono* (Florence, 1612) he gave a fairly good theory of Fluidstatics. It was, by the way, in this publication that he introduced the notion 'momento' (momentum).

If, he says, two bodies of absolutely equal weights (today we say masses) move with equal velocities, then they have the same 'power', or the same momentum. 'This is that property, that action, that power, by which the motive agency moves and the body resists.' I should, perhaps, add that the momentum of a body is defined today as the product of its mass and the linear velocity, while the moment of momentum, or angular momentum, of a body is the product of its moment of inertia and angular velocity. Thus, linear momentum is $m\bar{v}$, and angular momentum is $I\bar{\omega}$. Both linear and angular momentum are vector quantities. To underline the importance of Galileo's 'momento', let us also recall that, because of the relation of momentum to force, as set forth in Newton's laws, this is a fundamental concept in dynamics generally, in fluiddynamics particularly, therefore Galileo, too, occupies a prominent place in the list of creators of the science of fluid flows.

As I have already said, Galileo conjectured that the speed of a falling body at each instant might be proportional to the time that had elapsed since the body had been set free; but how was he to measure such short periods of time?

He first improved the water-clock in a very ingenious way, by letting the water drip into a receiver and then weighing the amount which had fallen with great accuracy; but the times to be measured were still uncomfortably short. Galileo accordingly slowed his experiments down, by substituting a slow roll down a gentle slope for the rapid vertical fall, in the belief that the same laws must govern both, as indeed is the case. He set up a gently sloping plank, some 12 yards in length, and made

† See *The Complete Works of Galileo*. Florence: 1908.

polished steel balls roll down a narrow groove cut into it. With this simple apparatus he was able to vertify his conjecture that the speed of fall increased uniformly with the time.

Galileo Galilei showed that the effect of force was not to *produce* motion, but to *change* motion, that is, to produce *acceleration*. Much later on, Sir Isaac Newton gave this conclusion the mathematical form of his famous Second Law; from then on, all branches of dynamics, including hydrodynamics, could grow on a sound philosophical basis.

This might be the right moment to make a note of special importance. The late Sir James Jeans wrote† that

> The Aristotelian had taught that all motion needed a force to maintain it, so that a *body on which no force acted* must stand at rest. In accordance with these ideas, Aristotle had himself introduced his Unmoved Mover, God Himself, to keep the planets in motion, while the mediaeval theologians had postulated relays of angles for the same purpose. It now appeared that to keep a body in motion it was only necessary *to leave it to itself; a body acted on by no force* would not in general stand at rest, but would move with uniform speed in a straight line, because there would be nothing to change its motion. (Author's italics.)

I think that these, and similar statements by other authors, are somewhat misleading, to say the least. For there are in the universe no bodies on which no force or forces act, and they never have existed, therefore no body can be 'left to itself', and no body acted on by no force could or will ever move with uniform or other speed.

Galileo knew, of course, that his rolling balls experienced both friction and air resistance, especially the latter, which he studied quite seriously. If there were no air resistance, he said, the trajectory of a ball leaving the horizontal plane would be a perfect parabola but 'the resistance of the air influences the path of the body'. He did not produce, however, a method for its determination and allowed himself to remark on another occasion that 'the air resistance is small enough to ignore it'. A certain Salviati criticized Galileo for this assumption, saying that, on the contrary, air resistance was 'very significant' and could 'make the parabola an unparabola'.

Already, Archimedes had disproved the teaching of the Aristotelian school that the shape of a body determined whether it would sink or float in water. Now Galileo proved experimentally that a body sinks not because of its shape, but because of its density relative to the density of the fluid in which it is immersed. Archimedes and Leonardo da Vinci not only used the notion 'pressure', but knew that every point of the

† *The Growth of Physical Science*. 1947.

wetted surface of a body in a fluid was under a definite pressure. Simon Stevinus was the first to distinguish between sticky (i.e. viscous) and non-sticky (i.e. ideal) fluids, and Galileo showed, both theoretically and experimentally, that the pressure at any point in an ideal fluid depended only on the 'head' of liquid above it (hence, by the way, comes the term 'head' in modern Fluidmechanics).

We are all familiar with the so-called siphon, which is, generally, a U-shaped tube through which liquid is made to flow upwards and carried across to a lower level (Figure 29). There is evidence that Archimedes

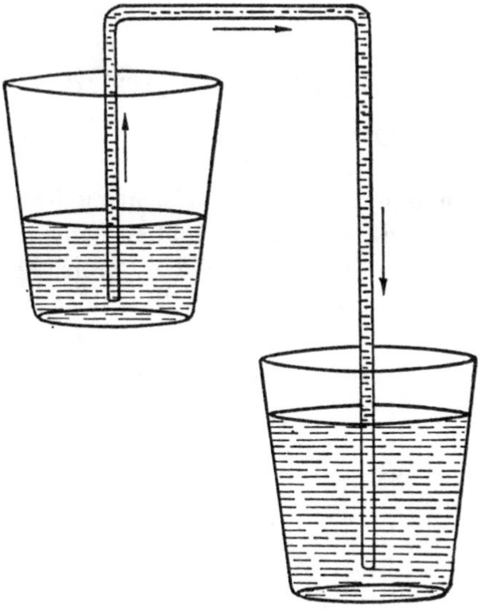

Fig. 29. The siphon principle

and/or Leonardo da Vinci were aware of this possibility. It seems incredible that it was unknown to Stevinus.

Galileo, however, not only knew of the siphon principle but also studied it, both theoretically and experimentally. And this led him to the inference that a small mass of liquid (contained in a narrow pipe) can keep in equilibrium a large mass of liquid (contained in a large vessel) because, as he put it, a slight lowering of the liquid level in the vessel entails a great increase of height in the tube.†

We are told‡ that his *Discorsi*, a book of Fluidstatics, was attacked by L. de Coulombe and V. di Grazia, and defended by Benedetto Castelli

† *The Complete Works of Galileo.* Italian National Edition, Florence: 1908.
‡ *A History of Mechanics.* Réne Dugas, Routledge and Kegan Paul, Ltd, London: 1955.

(1577–1644). The latter was the author of a treatise on the measurement of running water (*Della misura dell'acque correnti*, 1628); he used Leonardo's velocity-area law $AV =$ const and asserted its importance in all problems of fluid-flows.

Leonardo was, perhaps, the first to attempt to determine the weight of the air in relation to the other three 'elements' of nature, and to give the matter at least some thought. Galileo was probably the first actually to weigh a balloon filled with air, in the cool and hot state.

Evangelista Torricelli (1608-47) and Otto von Guericke (1602-86)

Toricelli was yet another outstanding Italian mathematician who contributed to the birth and formation of Fluidmechanics quite significantly. Among other things, he took up Galileo's projectile trajectories and developed them a step further. For example, Galileo made only a passing remark that two exactly similar bodies set in motion at the points A and B (Figure 30) with exactly equal velocities would move along the same

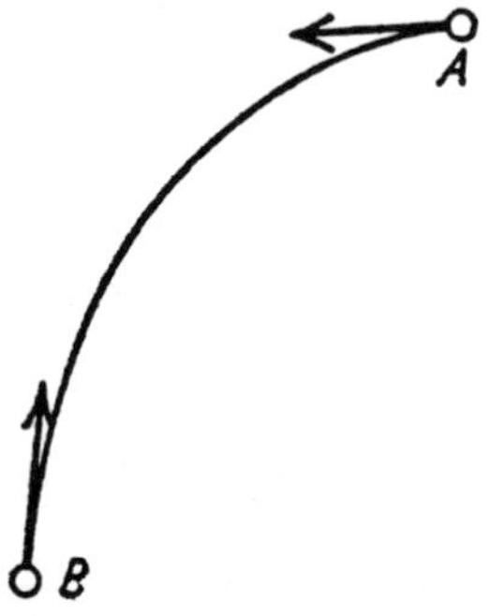

Fig. 30. Torricelli's parabola

parabolic trajectory. But Torricelli analysed the problem more fundamentally,† and gave it a solid theoretical proof.

It is known that Galileo was interested in Aristotle's and others' *horror vacui*. But it was his pupil and colleague Vincenzo Viviani (1622–1703) who worked out a method of investigation and carried it out in Florence in 1643. To understand what exactly Torricelli and Viviani did

† *A History of Mechanics*. Réne Dugas, Routledge and Kegan Paul, Ltd, London: 1955.

and why, I should like to advise you to perform a simple experiment of your own, in your own kitchen or bathroom. Fill an ordinary glass with water, then close it with the palm of your own hand; in that state turn it upside down, immerse it slightly in water (contained in anything), remove your hand, and you will observe two things: one, the water in the glass will not flow down, although it is higher than the water in which it is immersed; two, above the water in the glass there will be a little space without water (Figure 31).

Fig. 31. A simple experiment to demonstrate Torricelli's Vacuum

This space devoid of water is the so-called 'Torricelli's Emptiness', otherwise known as 'Torricelli's Vacuum'. Later on, Viviani replaced the cup by a bent glass tube, and water by mercury, which gave him a much better opportunity to measure the dimension of the vacuum. Now, what is holding the water in the cup and the mercury in the tube? The atmospheric pressure, of course. The greater this pressure, the higher the column of the water and/or of the mercury or, which is the same, the smaller the height of the vacuum. And so the experiments of Torricelli and Viviani led to the invention of the *manometer*, a very important instrument in Fluidmechanics.

The reader will see later on why I do not add that the experiment disproved the *horror vacui* of Aristotle and others.

Ten years or so earlier, but quite independently, the *horror vacui* was investigated also by Guericke in Magdeburg, Germany. His experiments were much more sophisticated, certainly more mechanical. You no doubt remember from your school physics textbook the following description: Guericke, the Burgomaster of Magdeburg, constructed two large hemispheres, attached them to each other, pumped out the air from the sphere thus formed, i.e. created a deep vacuum inside the sphere, and sixteen horses, four pairs on each hemisphere, were needed to pull the hemispheres apart. So, Guericke, too, proved that deep vacuum could exist and was possible: deep, but not absolute vacuum.

But Torricelli is also known in the history of Fluidmechanics as the inventor of hydrodynamics, a somewhat pompous title incommensurate with his actual contributions. Indeed, we have tried to put together all his works on the dynamics of fluid flows, but have failed to find enough *original* evidence to justify the title. There is, however, no doubt that he studied the characteristics of water discharge streams. Having made a narrow hole *a* near the bottom of a vessel filled with water (Figure 32),

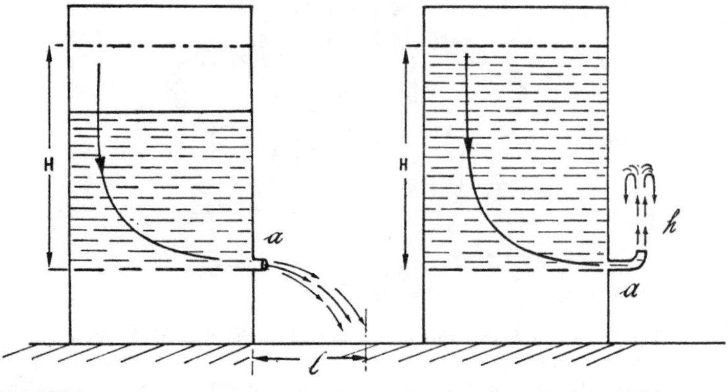

Fig. 32. Torricelli's scheme for the derivation of his celebrated 'Torricelli's Law'

and then, inserting into it orifices of various configurations, he noticed that, first, it was possible to 'strengthen' the discharge stream by means of an orifice; second, that for a given orifice the distance *l* depended entirely on the level *H* of the water in the vessel.

Torricelli then asked himself this question: the discharge stream is strong enough to make a distance *l*; but to what height *h* could it rise, if it were forced to follow upwards? No doubt, he attempted to answer the question theoretically, and he probably determined the weight of the liquid above the orifice level as 'the force that pushes the stream through the hole'. He then experimented with actual upward streams and established, to his surprise, that always $h < H$, or $H - h = \triangle h > 0$.

There were, he said quite rightly, two reasons for this: first, the frictional resistance of the orifice, and, second, air resistance to the stream outside the orifice. Then, neglecting the latter, he recalled two facts established earlier: first, the distance of free fall $s = h = \text{const} \times t^2 = gt^2/2$, and, second, the velocity of free fall $v = gt$. From these he deduced the famous Torricelli's Law,

$$s = h = \frac{g}{2}\frac{v^2}{g^2} = \frac{v^2}{2g}, \quad \boxed{v = \sqrt{2gh}},$$

which was at the time one of the greatest milestones, if not the greatest, in the history of Fluidmechanics. We shall meet the formula more than once (in the section dealing with Daniel Bernoulli, for instance).

Blaise Pascal (1623-62)

Pascal, one of the great French scientists and philosophers, carried on experiments on the equilibrium of fluids, and differences in barometric pressure at different altitudes. He repeated Torricelli's experiment, using red wine and a glass tube and explained the siphon more fully. A brilliant Christian apologist and outstanding scientist, he tried to build a reliable bridge between supernatural dogmas and scientific discoveries, but never succeeded. In his *The Equilibrium of Liquids and the Weight of the Mass of the Air* (1663), Pascal not only embraced all the hydro- and aerostatic problems studied by his predecessors, but extended and developed them into a coherent theory of Fluidstatics.† Problems studied by him were: how the weight of liquid varies with depth; the equilibrium of liquids; equilibrium between a liquid and a solid; bodies wholly immersed in water; immersed compressible bodies; animals in water, etc.

Having thus established basic laws for water at rest, Pascal could, and did, easily proceed to the study of the atmosphere. He knew that the 'heaviness' of the latter had been demonstrated a long time ago, by the Milanese mathematician and philosopher of nature, Girolamo Cardano (1501–76), but he wanted to determine its actual weight. And while

† *Collection of Pascal's Works*. (Translated into English by H. B. and A. G. H. Spiers), Columbia University Press, New York: 1937.

solving this and associated problems, Pascal formulated almost all the basic laws of aerostatics, e.g. that (1) since every part of the air has weight (mass), it follows that its whole body also has weight (mass); (2) since the mass of the air covers the entire face of the earth, its weight presses upon the earth everywhere; (3) because of the hydrostatic law, the higher parts of the earth, such as summits of mountains, experience less air pressure than the lowlands; (4) bodies in the air are pressed on all sides.

He then draws attention to everyday experience: when all the apertures of a pair of bellows are closed, why are they difficult to open; when two polished surfaces are laid one upon another, why are they as difficult to separate as if they had been glued together; when a syringe is dipped in water and the piston drawn back, why does the water follow it as if adhering to the plunger, etc.?

'The basic reason for all these phenomena is the weight of the air,' wrote Pascal, 'although it had hitherto been ascribed to the horror of a vacuum. . . .' That is, he knew that a kind of vacuum was indeed possible. But he also knew that the invisible power of the atmosphere could do 'wonders of pressure'. One could, perhaps, say figuratively that the day the learned and learning world realized the role played by the atmospheric pressure, the gates on the road to Fluidstatics were wide open.

One of the conclusions of Pascal was his theorem that a given volume of fluid weighs according to the depth (the atmospheric altitude) at which it is taken. The other interesting conclusion was that, as he himself put it, 'a body never moves because of its weight – it moves only when its centre of gravity moves'. The first conclusion led man to the establishment of the laws of change of the physical properties of the atmosphere with altitude; the second became the ABC of hydraulic mechanics and machines of all kinds.

Indeed, he says, pistons *A* and *B* (Figure 33) are in equilibrium because their common centre of gravity is at a point which divides the

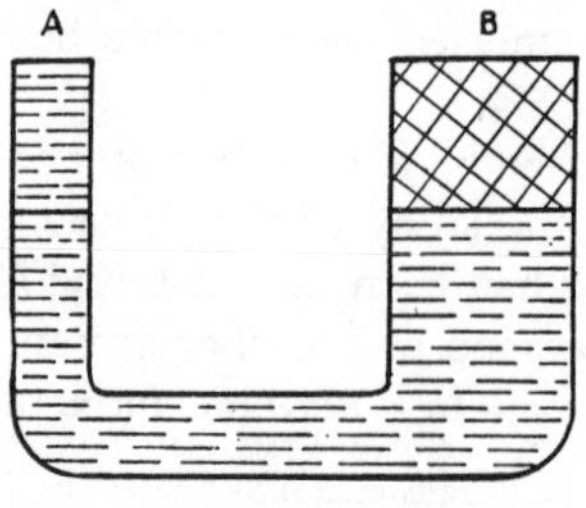

Fig. 33. Pascal's scheme of hydrostatic equilibrium

line passing through their centres of gravity in proportion to their weights. But if they move, their paths will be related to each other as their reciprocal weights, and their new common centre of gravity will be found in exactly the same position as before.

The engineering consequences of this simple principle proved to be revolutionary. It had been known, even before Stevinus and Pascal, that the pressure of a fluid is normal to the surface on which it acts. Secondly, Stevinus and especially Pascal showed that pressure applied to the surface of a fluid is almost instantly transmitted to all parts of the fluid. Thirdly, Stevinus assumed, and Pascal proved, that the pressure at any point within a fluid is the same in all directions, and depends only on the depth.

The pressure of a fluid is measured by the force (kilograms) exerted on a unit of area (square metres), i.e. p = force/area. It was found that if a closed vessel is filled with water, and if A and B are two equal openings in the top of the vessel (Figure 34), closed by pistons, any pressures

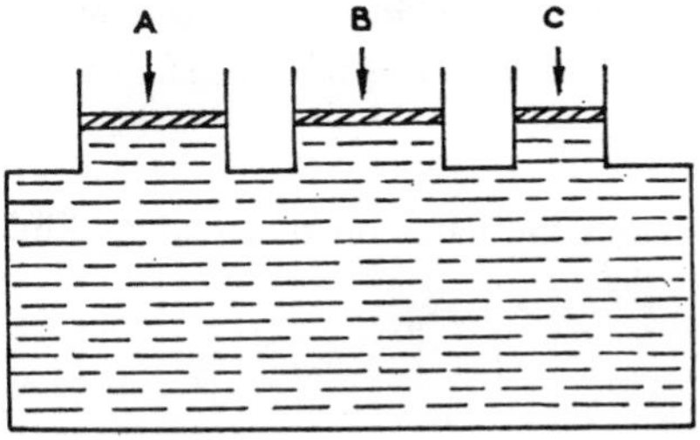

Fig. 34. The principal of proportional hydrostatic pressures

applied at A must be counteracted by an equal pressure at B to prevent its being forced out; and that if C be a piston of a different size, the pressure applied to it must be related to the pressure on A as the ratio of the area of C is to that of A, and that this was the case whether the piston B existed or not.† More generally, if a vessel of any shape has several openings closed by pistons, kept at rest by suitable forces, any additional force P applied to one piston will require the application, to all the other pistons, of additional forces which have the same ratio to P as the respective pistons have to that of the piston to which P is applied.

From Stevinus onwards, everyone concerned with problems of this nature was anxious to understand the mechanism of transmission of pressure in fluids. Different people offered different explanations and proofs, which resulted in the following one, usually attributed to Pascal.

† *Elementary Hydrostatics*. W. H. Besant, Deighton, Bell & Co., Cambridge: 1873.

Let A and B be two points in a fluid at rest, and about the straight line AB as axis describe a cylinder having plane ends perpendicular to AB, and imagine that this cylinder is solid (Figure 35). The equilibrium of

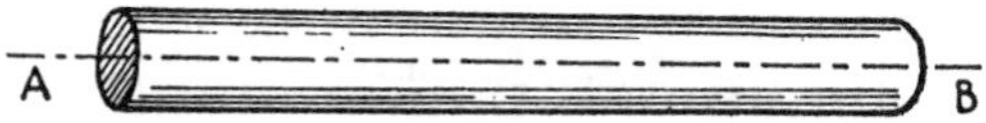

Fig. 35. Imaginary fluid column

this cylinder is maintained by the fluid pressure on its ends, which are parallel to the axis, by the fluid pressures on its curved surface, which are perpendicular to its axis, and by its weight, which is vertical. Now resolving along AB, the difference of the pressures at A and B must be equal to the resolved part of the weight in the direction BA, and the weight remaining the same, any change of pressure at A involves the same change at B. Moreover, if fluid be contained in a vessel of any shape, and the straight line AB does not lie entirely in the fluid, the two points may be connected by series of lines such as $ACDB$ (Figure 36),

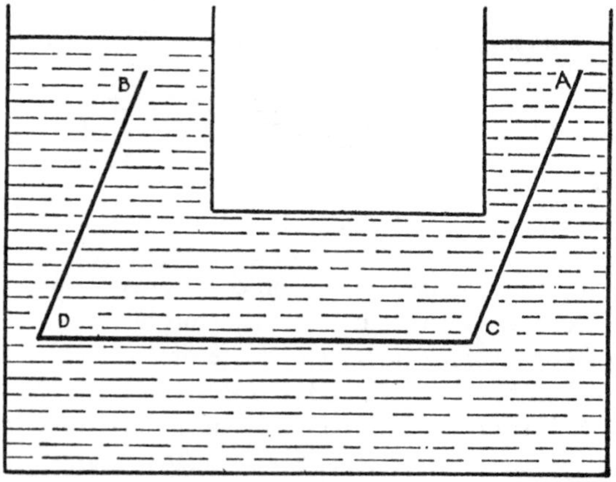

Fig. 36. A scheme to prove the distribution of hydrostatic pressures

and any change of pressure at A produces an equal change at C, and that change produces an equal change at D, which, in turn, produces the same change at B, and vice versa.

But both Stevinus and Pascal, and many others, had still to prove the equality of pressure in all directions, at any point, within the fluid. The proof was provided by Pascal. He imagined an infinitesimal prism of fluid (Figures 37 and 38) in equilibrium. Taking d for the length of the

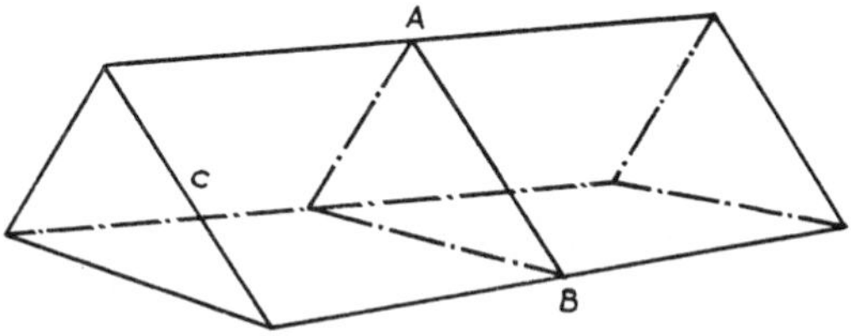

Fig. 37. Resultant pressure forces acting on the faces of an infinitesimal fluid prism

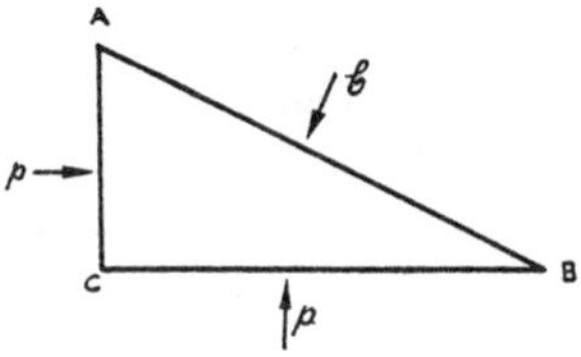

Fig. 38. An infinitesimal prism of fluid

prism, a, b, c for the sides of the triangle, w for the weight of a unit volume of the fluid, and p, p', p'' for the pressures on the sides AC, CB and BA, he stated, quite rightly, that the resultant pressure forces were $(p.bd)$, $(p'ad)$ and $(p''dc)$, and the weight $= \frac{1}{2}\, ab.dw$.

Hence, resolving vertically and horizontally,

$$\frac{1}{2}\, abw = p'a - p''c \cos B,$$
$$pb = p''c \sin B;$$

but $a = c.\cos B$, $b = c.\sin B$, therefore $p = p''$, and $p' - p'' = \frac{1}{2}\, b.w$. He then supposed that the sides a and b diminished indefinitely, in which case p, p' and p'' had to become pressures in different directions at the point C. He thus proved that $p' = p''$ and, consequently, $p = p' = p''$, i.e. the three pressures are equal at the point C. By turning the prism round AC and changing the angles A and B, he showed that the proposition was true for all directions.

All these were fundamental developments in Fluidstatics. But Fluid-mechanics, as a whole, was still to emerge; the science of motion of fluids under the action of forces still did not exist, although Torricelli made an important step in that direction.

Sir Isaac Newton (1642-1727)

It is impossible to pin down any particular Newtonian contribution to Fluidmechanics; for almost all the fundamental fluidmechanic concepts are built upon Newton's basic laws.

His second law introduced a new conception of force: as that which changes momentum, the latter being defined as the mass of a body multiplied by its speed of motion; and mass was defined by him as 'the volume of a body multiplied by its density'. Let me say right away that there are in nature no Newtonian (concentrated) forces; but Newton's definitions met, and continue to meet, all the basic needs of classical mechanics, therefore we shall 'tolerate' them for the time being without comment (see the last section of this book).

The second point of major historic importance is this. We have seen that problems of gravitation, inertia and motion of bodies were formulated before Newton. But none of them emerged in that simple, beautifully finalized, form as Newton's Laws of Motion. His *Philosophiæ Naturalis Principia Mathematica* became the hinge of all the levers of Mechanics—'the greatest production of the human mind' (Lagrange). As Laplace put it, Newton's contribution to scientific knowledge surpassed everything the entire history of science had achieved before, and *Principia* will always shine as the star of human genius.

If we condense Newton's Mechanics, the following laws emerge as its essential roots:

A material body does not alter its motion in any way, except under the action of a force applied to it; a material body at rest remains at rest, or in uniform motion – it continues to move in the same direction, with the same speed, unless a force is impressed upon it; the time-rate of change of momentum is proportional to the force which causes it; to every action of a force there exists a counter action, or a reaction. In a still more condensed, and popular, form these are reduced to the following Three Laws of Motion: (1) every material particle continues in its state of rest or of uniform motion in a straight line except insofar as it is compelled by force to change that state; (2) the time-rate of change of momentum ($m\bar{v}$) is proportional to the motive force and takes place in the direction of the straight line in which the force acts; and (3) the interaction between two particles is represented by two forces equal in magnitude but oppositely directed along the line joining the particles.

We shall see in due course that these laws, in spite of their profound importance, are open to philosophical criticism. But even these criticisms require the knowledge of . . . the same laws!, especially of the second law, which has the mathematical form

$$\boxed{\bar{F} = \frac{d}{dt}(m\bar{v}) = m\frac{d\bar{v}}{dt} = m.\bar{a} = \frac{W}{g}\bar{a}},$$

where $\bar{F}$ = force, m = mass, $\bar{v}$ = velocity, $\bar{a}$ = acceleration, W = weight, t = time, d = symbol of differentiation.

Sir Isaac Newton postulated that every fluid in nature consists of perfectly spherical elastic particles at equal distances from each other. Thus, according to Newton's first theory, a solid body moving in such a medium imparts momentum to all particles it meets on its way. The particles do not communicate their motion to neighbouring particles.

Starting from here, Newton tried to develop a theory of fluidmechanic resistance and produced interesting results. For a cylinder, he found that the resistance was equal to the weight of a cylinder of fluid of the same base as the solid, and whose height was twice that from which a heavy body would have to fall in order to acquire the velocity with which the solid moved. The resistance of a sphere appeared to be one-half the resistance of the cylinder, under the same conditions.

But it was Newton himself who soon disproved these conclusions. Assuming now that a fluid stream is not a stream of isolated balls lined up in perfect order, but a continuous chain of particles, he studied the effect of such a stream upon a curved surface and arrived at the new conclusion that the resistance experienced by a cylinder in translational motion was equal to the weight of a cylinder of fluid whose base is the same as that of the solid and whose weight is half that from which a heavy body would have to fall to acquire the velocity with which the solid moves in the fluid, i.e. four times less than by the first theory. He also found that the length of the cylinder did not affect the resistance – we shall return to this conclusion later.

The essence of Newton's theory is that molecules, or particles, are assumed to move in straight lines until they strike the body surface. When this happens, they lose the component of their momentum normal to the body. If we take, for example, an inclined flat plate (Figure 39) we shall have

$$\boxed{\begin{aligned} R &= (mv)_n = (\rho.A.v_\infty v)_n = \rho A_n v_\infty v_n = \rho A \sin\alpha\, v_\infty v_\infty \sin\alpha = \\ &= \rho A v_\infty^2 \sin^2\alpha \end{aligned}},$$

where ρ is the mass density of the fluid, A the area of the plate, and α its angle of incidence. This is, then, the so-called Newton's Sine-Square Law of air resistance. But, in fact, it cannot be found in Newton's work; it was deduced by other investigators.†

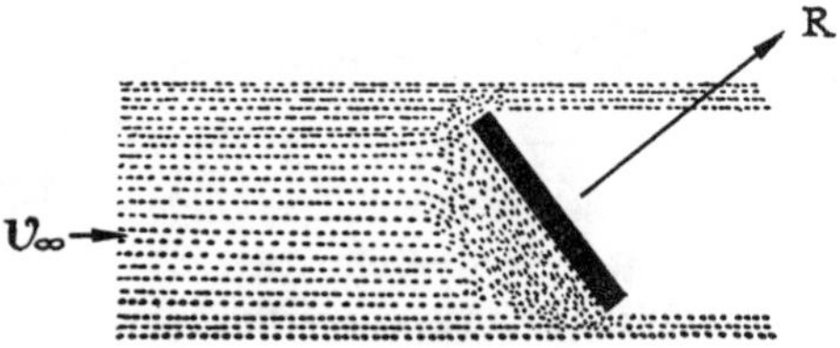

Fig. 39. Newton's model of flow past an inclined plate

We have so far been surveying the main milestones in the history of mechanics of ideal, inviscid, fluids. But this does not mean that the existence of friction in real fluids was unknown to the developers of Fluidmechanics.

Indeed, Michel Angelo (1475–1564), more widely known as Michelangelo (Buonarroti), the great Italian sculptor, painter, architect and poet of the High Renaissance, who was the author of many monumental hydraulic and other engineering projects in Rome, observed that the velocity of a water flow is greater in the centre of the flow than near the banks. This suggests that he knew of the existence of fluid friction. Leonardo da Vinci, in turn, went so far as to define the main parameters affecting the resistance met by a solid body moving in a real fluid. René Descartes (1596–1650), or Renatus Cartesius in Latin (hence the name Cartesian Coordinates), a French scientist and philosopher, who in his *La Géometrie*‡ established the Cartesian system of mathematical certitude, also studied the problem of friction between two liquid layers. Then there are Torricelli and Viviani who even tried to establish experimentally a relationship between the kinematic and frictional characteristics of water jets, while d'Alembert came to the conclusion that it was impossible to obtain reliable results without the help of experiments. 'I must confess', he wrote, 'that I do not know how the resistance of fluids can be explained theoretically, because theory gives zero resistance. . . .'

Nevertheless, the determination of the force of fluid resistance had to wait until the end of the XVII century, when Newton in England and Guillelmini in Italy, independently of each other, published their famous works. In his *Della natura de fiumi* (Roma, 1697), Guillelmini made a

† *Aerodynamics*. Theodore von Karman, Cornell University Press, Ithaca: 1954.
‡ See, for instance, *Philosophical Writings*. Descartes (selected and translated into English by Norman Kemp Smith), Random House, New York: 1958.

definite attempt to analyse the physical nature, and to establish the mathematical structure, of friction between fluids and solid surfaces.† But his effort was superseded, both in time and essence, by Sir Isaac Newton.‡

Newton, incidentally, wrote that a fluid is a body whose particles can move relative to one another under the action of *any* force, which suggested the absence of friction. In the same book, however, he not only accepts its existence, but determines its force: even Newton was not free of contradictions!

If portions of a mass of fluid are caused to move relative to one another, says Newton's theory, the motion gradually subsides unless sustained by external forces. Conversely, if a portion of a mass of fluid is kept moving, the motion gradually communicates itself to the rest of the fluid. These effects, observed generally long before he was born, were ascribed by him to a *defectus lubricitatis*, that is, to a lack of slipperiness, or internal friction, or viscosity in modern terminology. The corresponding foreign terms are: *frottement interieur* and *viscosite* in French; *innere Reibung* and *Viskosität* or *Zähigkeit*, in German; *vnutrenneye treniye* and *vyazkost*, in Russian.

If A and B (Figure 40) are two particles of a viscous flow sliding one

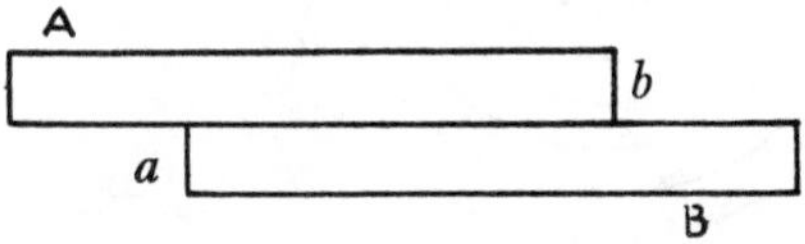

Fig. 40. Two fluid layers sliding over each other

over another, then there exists friction, or viscous resistance, along the surface *ab*. The force of this resistance is known as the *shear force*, while the shear force per unit area of friction is known as the *shear stress*.

The magnitude of the shear stress depends on the speed with which the two layers slide one over another:

$$\boxed{\tau = \frac{F}{A} = \mu \frac{dv}{dy}},$$

where μ is the absolute or dynamic viscosity coefficient, such that

$$\boxed{\mu = \rho\nu},$$

† *Traite d'hydrodinamique*. Bossu, t. 2.
‡ *Philosophiae naturalis principia, Mathematica*, Book II. Newton, 1687.

ν being the so-called 'kinematic viscosity', and ρ the mass density of the fluid.

With reference to Figure 41, Newton's actual description is as follows:

'If a solid cylinder infinitely long, in an uniform and infinite fluid, revolves with an uniform motion about an axis given in position, and the fluid be forced round by only this impulse of the cylinder, and every part of the fluid continues uniformly in its motion: I say, that the periodic times of the parts of the fluid are as their distances from the axis of the cylinder.

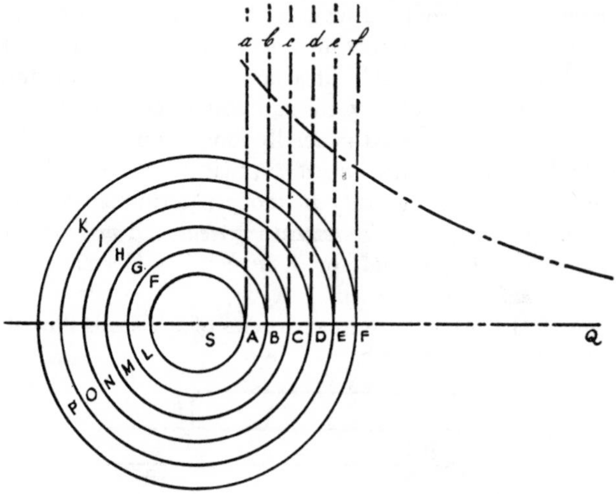

Fig. 41. Newton's scheme of fluid friction

Let *AFL* be a cylinder turning uniformly about the axis *S*, and let the concentric circles *BGM*, *CHN*, *DIO*, *EKP*, etc., divide the fluid into innumerable concentric cylindric solid orbs of the same thickness. Then, because the fluid is homogeneous, the impressions which the contiguous orbs make upon each other will be (by the Hypothesis) as their translations from each other, and as the contiguous surfaces upon which the impressions are made. If the impression made upon any orb be greater or less on its concave than on its convex side, the stronger impression will prevail, and will either accelerate or retard the motion of the orb, according as it agrees with, or is contrary to, the motion of the same. Therefore, that every orb may continue uniformly in its motion, the impressions made on both sides must be equal and their directions contrary. Therefore since the impressions are as the contiguous surfaces, and as their translations from one another, the translations will be inversely as the surfaces, that is, inversely as the distances of the surfaces from the axis. But the differences of the angular motions about the axis are as those translations applied to the distances, or directly as the translations and inversely as the distances; that is, joining these ratios together, inversely as the squares of the distances. Therefore, if there be erected the lines *Aa*, *Bb*, *Cc*, *Dd*, *Ee*, etc., perpendicular to the several parts of the infinite right line *SABCDEQ*, and inversely proportional to the squares of *SA*, *SB*, *SC*, *SD*, *SE*, etc., and through the extremities of those perpendiculars there be supposed to pass a hyperbolic curve, the sums of the differences, that is, the whole angular motions, will be as the correspondent sums of the lines *Aa*, *Bb*, *Cc*, *Dd*, *Ee*, that is (if to constitute a medium uniformly fluid the number of the orbs be increased and their breadth diminished *in infinitum*) as the hyperbolic areas *AaQ*, *BbQ*, *CcQ*, *DdQ*, *EeQ*, etc., analogous to the sums; and the times, inversely proportional to the angular motions, will be also inversely proportional to those areas. Therefore, the periodic time of any particle, as *D*, is inversely as the area *DdQ*, that is (as appears from the known methods of quadratures of curves), directly as the distance *SD*.

Q.E.D.

COR. (corollary) I. Hence the angular motions of the particles of the fluid are inversely as their distances from the axis of the cylinder, and the absolute velocities are equal.

COR. II. If a fluid be contained in a cylindric vessel of an infinite length, and contain another cylinder within, and both the cylinders revolve about one common axis, and the times of their revolutions be as their semidiameters, and every part of the fluid continues in its motion, the periodic times of the several parts will be as the distances from the axis of the cylinders.

COR. III. If there be added or taken away any common quality of angular motion from the cylinder and fluid moving in this manner, yet because this new motion will not alter the mutual attrition of the parts of the fluid, the motion of the parts among themselves will not be changed; for the translations of the parts from one another depend upon the attrition. Any part will continue in that motion, which, by the attrition made on both sides with contrary directions, is no more accelerated than it is retarded.

COR. IV. Therefore, if there be taken away from this whole system of the cylinders and the fluid all the angular motion of the outward cylinder, we shall have the motion of the fluid in a quiescent cylinder.

COR. V. Therefore, if the fluid and outward cylinder are at rest, and the inward cylinder revolve uniformly, there will be communicated a circular motion to the fluid, which will be propagated by degrees through the whole fluid; and will go on continually increasing, till such time as the several parts of the fluid acquire the motion determined in Cor. IV.

COR. VI. And because the fluid endeavours to propagate its motion still further, its impulse will carry the outmost cylinder also about with it, unless the cylinder be forcibly held back; and accelerate its motion till the period times of both cylinders become equal with each other. But if the outward cylinder be forcibly held fast, it will make an effort to retard the motion of the fluid; and unless the inward cylinder preserve that motion by means of some external force impressed thereon, it will make it cease by degrees. All these things will be found true by making the experiment in deep standing water.

The role of temperature in fluid friction – viscosity – was neglected by Newton, but studied subsequently by Du Buat, Girard and others, and especially by Poiseuille. They showed that the force of friction decreases with the increase of temperature, and this, in turn, brought forward yet another question: what is the physical nature of viscosity?

Navier, first (150 years after Newton!) introduced μ into the general equations of motion (see p. 88), and Poisson ascribed fluid friction to the change of the 'repulsive forces' of fluid particles, a concept unacceptable to modern Physics. Saint-Venant, Kleitz, Helmholtz, Meyer, Kirchhoff and others advanced yet another theory, according to which rapid displacements of particles in a flowing fluid create in it forces which are proportional to the relative velocities of the layers of the fluid.

Daniel Bernoulli (1700-82)

Just as a great river is fed by small streams, some even barely noticeable at its source and along its banks, so science and technology proceeds from small individual contributions until it becomes an ever-increasing flow of knowledge and techniques. This big river of Fluidmechanics is closely associated with Daniel Bernoulli, the author of the first textbook in the field.† Reading and re-reading it, I entirely agree with its author that his theory was 'novel, because it considers both the pressure and the motion of fluids'. Since the book is not readily available to the average reader, I should like to give here some details of Bernoulli's methods and techniques.

Chapter 13 of the book is called *Hydrodynamica*. Let there be a large vessel *AFEB*, (Figure 42), begins Bernoulli, which must constantly be

† *Hydrodynamica, sive de viribus et motibus fluidorum commentarii*. Berlin (written in St. Petersburg): 1738.

kept full of water. Let the vessel have a horizontal cylindrical pipe *BD*, with a hole *O* at its end, through which the water is ejected with a steady velocity. It is required to determine the pressure of the water on the walls of the pipe *BD*.

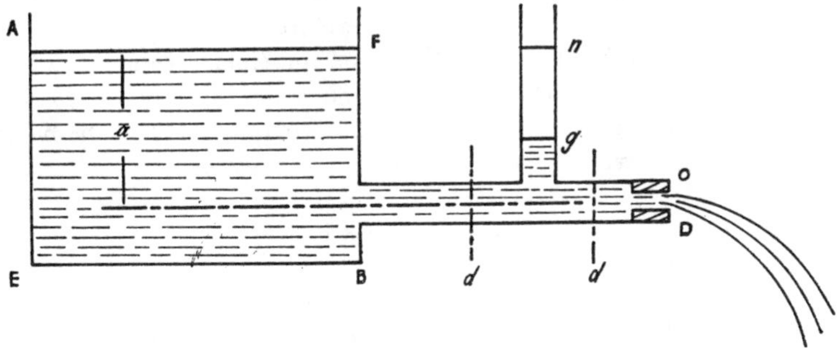

Fig. 42. Bernoulli's historic scheme of discharge of water, from which he derived his famous 'Bernoulli's Law'

Solution: let the height of the surface of the water above the hole *O* be equal to a; then the velocity of the water flowing out at *O*, if the first instants of the outflow are excluded, should be considered as steady and equal to $\sqrt{a}$, for we accept that the vessel is constantly kept full. And if one assumes that the ratio of the cross-section of the pipe and of its hole is equal to $n/1$, then the velocity of the water in the pipe will be equal to $\sqrt{a}/n$. But if orifice *OD* were missing entirely, then the limiting velocity of water in the same pipe would be equal to $\sqrt{a}$, which is greater than $\sqrt{a}/n$. Thus, water in the pipe tends to accelerate, but meets the resistance offered by the end *OD*. These obstacles and counter-actions compress the water; the compression is transmitted to the walls of the pipe which experience, therefore, the same increased pressure. Thus, it is clear that the pressure on the walls is proportional to the acceleration or to the increase of velocity, which would be gained by water, if the obstacle were to disappear instantaneously and the water to flow right into the air.

So the problem is to determine the acceleration which would be gained by the particle *dd*, if the pipe *BD* were at some instant cut at *d*, the water continuing to flow through *O*; this is precisely the pressure of flowing water upon particle *dd* taken on the wall of the pipe. To find its value, consider the whole vessel *ABddD* and find for it the acceleration in the close neighbourhood of the outflowing water particle, which has velocity $\sqrt{a}/n$.

. . . Let v be a variable velocity in the pipe *Bd*. The cross-section of the pipe, as before, is n, and $Bd = c$, $dd = x$. At B, there is an equal particle ready to enter the pipe at the instant when particle dd leaves it. But when the particle at B, whose mass is equal to $n.dx$, enters the pipe, it acquires the velocity v and also the living force (i.e. kinetic energy) nv^2dx. Since vessel AB is infinitely large, the particle at B has been in a state of rest before entering the pipe, therefore nv^2dx is an entirely new force. We must add to this living force the increment of the living force gained by water at Bb, while particle dd flows out, namely $2ncvdv$. This sum corresponds to the real descent of the particle ndx from the height BE, i.e. a. Thus, we have $nv^2dx + 2ncvdv = nadx$, or

$$\boxed{\frac{vdv}{dx} = \frac{a - v^2}{2c}}$$

But, in any flow, the velocity increment dv is proportional to pressure multiplied by time, which in this case is equal to dx/v. Therefore, in our case, pressure experienced by particle dd is proportional to vdv/dx, that is, to $(a - v^2)/2c$.

At the instant when the pipe is cut, $v = \sqrt{a}/n$ or $v^2 = a/n^2$; this expression should be substituted in the right-hand side of the equation, which then becomes $(n^2 - 1)a/2n^2c$. And the latter represents a quantity proportional to the pressure of the water on the portion ac of the tube, whatever the cross-sections of the tube and of the orifice at its end.

So, if the pressure of the water is determined for one case, it is determined also for all the other cases, and this is so when the orifice is infinitely small or when n is infinitely large compared with unity; because it is self-evident that, in such a case, the water exerts its entire pressure corresponding to a; this pressure we call a. But when n is infinitely large, then unity is vanishingly small compared with n^2, and the quantity, which is proportional to pressure, appears to be equal to $a/2c$. Thus, if we wish to know what the pressure will be at any n, we should ask this question: if a corresponds to quantity $a/2c$, what pressure will correspond to quantity $(n^2 - 1)a/2n^2c$? In this way, one establishes that the answer is $(n^2 - 1)a/n^2$.

It follows from the disappearance of c from the calculation that all parts of the tube, those near the vessel AB as well as the more distant ones, experience the same pressure from the flowing water, which is less than that on the bottom EB. The greater the orifice O, the greater is this pressure difference. And the walls experience no pressure at all, when obstacle OD is missing, because, in this case, the water flows out through a full opening.

If we make somewhere in the wall of the tube a hole very small compared with O, water will shoot out through it with a velocity sufficient to elevate it (the water) to the height $(n^2a - a)/n^2$, if there are no obstacles in its way. . . .

The next problem Bernoulli analyses and solves is this: determine the pressure of water flowing with a constant velocity in a tube of any form. The solution obtained is

$$\boxed{\frac{vdv}{dx} = \frac{a - v^2}{2a}},$$

where a is a constant number whose value depends on the geometry ot the tube. If the actual pressure is $(a - b)$, b being the height corresponding to the actual v, and if $(a - b) < O$, then pressure becomes suction, i.e. the walls of the tube experience pressure from outside.

The rest of the Chapter (24 more pages) is given in the same typically Bernoulli style. The two enframed formulae are the two slightly different forms of the actual Bernoulli Equation. They appear on a number of pages of the textbook in modified forms. And so we find ourselves face to face with one of the most fascinating questions in the history of Fluidmechanics: who was, then, the author of the famous 'Bernoulli Equation' which appears in every textbook of Hydrodynamics,

$$\boxed{\frac{\rho v^2}{2} + p = \text{const}}\ ?$$

Indeed, this formula has no direct resemblance to the above two equations by Bernoulli. It is true that he was the first to undertake a fundamental study of the $p = p(v)$ or $v = v(p)$ interdependence, and the first to conclude that an increase in v leads to a definite decrease in p, and vice-versa, but certainly not in the form of the formula which is ascribed to him. This 'mystery', however, is dispelled by a close examination of Euler's equations of motion.

Leonhard Euler (1707-83)

Leonhard Euler was not a contributor to, but the founder of, Fluidmechanics, its mathematical architect, its great river.

Let us recall that geometry is a branch of mathematics which treats the shape and size of things; while Fluidmechanics is the science of motion (and equilibrium) of bodies of deformable (and variable) shapes, under the action of forces. When one analyses these two definitions, it becomes clear that *some* theorems and axioms of geometry do not meet the philosophical and physical needs of mechanics generally, and of Fluidmechanics in particular; and it is difficult to imagine that Euler's genius could have been unaware of this.

For example, a point is usually defined as an element of geometry which has position but no extension; a line is defined as a path traced out by a point in motion; and motion is defined as a change of position in space. But motion and matter cannot be divorced. A point that has no extension lacks volume and, consequently, mass, therefore is nothing; and nothing can have neither path nor momentum, or motion.

Euler was, perhaps, the first to overcome this fundamental contradiction, by means of the introduction of his historic 'fluid particle', thus giving Fluidmechanics a powerful instrument of physical and mathematical analysis. A fluid particle is imagined as an infinitesimal body, small enough to be treated mathematically as a point, but large enough to possess such physical properties as volume, mass, density, inertia, etc. Like Newton, Euler defined mass as the product of volume by the *mass density* of the fluid which occupied the volume. Hence, the now classical definition of mass density: it is the amount, or quantity, of fluid per unit volume. Mathematically,

$$\boxed{\rho = \frac{dm}{dV} = \frac{M}{V}} \quad (\text{kg. sec}^2/\text{m}^4)$$

But since, by Newton's Second Law, the weight W of any physical body $W = m.g$, it follows that

$$\rho = \frac{W}{gV} = \frac{\gamma}{g}, \text{ or } \boxed{\gamma = \rho g} \quad (\text{kg/m}^3).$$

So, the fluid particle concept became meaningful, logical, powerful. From then on, everyone knew that a fluid particle was not a mathematical, but a *physical* point possessing volume, weight, mass, densities, specific heats, etc. But in what shape should it be imagined? and how should its motion be determined?

Again, Euler produced the most beautiful answers, directly or by implication. Let us imagine a continuous curved line l, within a fluid flow, at any given instance of time tangential to velocity vectors of all fluid particles through which it passes, the so-called stream-line. The word *tangential* implies that anywhere along the stream-line the velocity vector is parallel to the portion of l where it acts. Euler exploited this fact in a somewhat complicated way; but if we apply to it the theorem (of vectorial mechanics) that the vector product of two parallel vectors is zero, we have:

$$\bar{v} \times dl = \begin{vmatrix} \bar{i}, & \bar{j}, & \bar{k} \\ v_x, & v_y, & v_z \\ dl_x, & dl_y, & dl_z \end{vmatrix} = 0$$

or, since the unit vectors $\bar{i} \neq 0, \bar{j} \neq 0$ and $\bar{k} \neq 0$, and denoting $dl_x = dx$, $dl_y = dy$, $dl_z = dz$,

$$\boxed{\frac{dx}{v_x} = \frac{dy}{v_y} = \frac{dz}{v_z}}$$

This is, then, Euler's classical differential equation of the stream-line, one of the golden keys to the mysteries of Fluidmechanics. It answers the second of the above two questions and, moreover, leads to a number of further ideas and concepts. For example, we can imagine, within the flow, a stream-tube composed of stream-lines (Figure 43). Now, since

Fig. 43. The imagery stream tube

the velocity vectors are parallel to the stream-lines at the points of their action, and since the walls of the stream-tube are composed of stream-lines, it follows that no amount of fluid enters or leaves the imaginary tube through the walls. Therefore, the amount (mass) of fluid entering through cross-section (I) per unit time must be exactly equal to the amount (mass) of fluid leaving the tube through cross-section (II) per unit time. That is, there can be in the stream-tube neither an accumulation nor a loss of mass, i.e. $m_1 - m_2 = dm = 0$. But $dm = \rho dV$.

What should the configuration of dV be? The general answer is: any configuration. Euler chose, however, an infinitesimal parallelopiped with dx, dy, dz as its sides (Figure 44) and, consequently, with $dV = dxdydz$ as its volume. Thus, the condition of 'no accumulation and no loss of mass' in the stream-tube assumed the form $dm = \rho dV = \rho dxdydz = 0$. And the integration of this expression produced yet another typically Eulerian historic formula,

$$\boxed{m = \underset{(V)}{\iiint} \rho dxdydz + \text{const}},$$

which represents, in fact, the law of conservation of mass in fluid flows, and remains fundamental in every branch of Fluidmechanics.

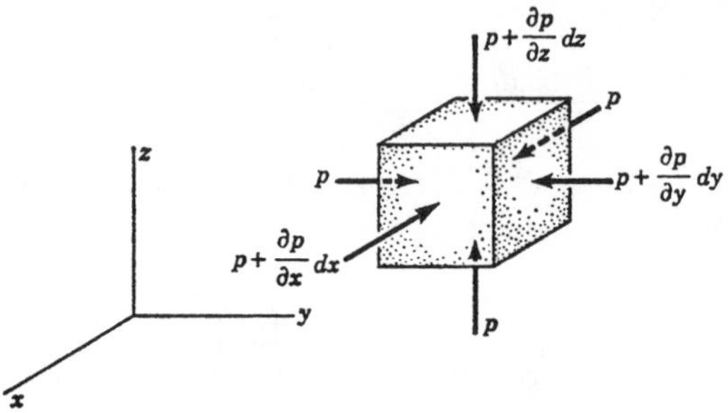

Fig. 44. Euler's infinitesimal fluid parallelopiped

Any magnitude which has size, in the ordinary algebraic sense of the word, as well as direction in space, is a *vector*; velocity, acceleration and force are examples of vectors. The common algebraic magnitudes, which have nothing to do with direction in space and have no directional properties, but are each determined completely by a single (real) number, are *scalars*; mass and temperature are typical examples of *scalars*. Leonhard Euler used in his analysis both vectors and scalars, but without calling them so. All his mathematical operations appeared in the Cartesian (rectangular) components. But they were, of course, the components of vectors. For if R_x, R_y, R_z be generalized rectangular components, then $\bar{i}R_x + \bar{j}R_y + \bar{k}R_z = \bar{R}$, the latter being the generalized vector, and $\bar{i}, \bar{j}, \bar{k}$ being its unit vectors.

Now, since Euler's hydrodynamic theory was based on the mass conservation and continuity concepts (actually, these are two modes of one and the same thing), it follows that R_x, R_y, and R_z (therefore also $\bar{R}$), whatever they may represent, must be continuous functions of space (co-ordinates) and time. Which leads us to important definitions. First, fluid particles in motion have weights, accelerations, velocities, masses, and so on. The space occupied by a flow – by particles in motion, that is – is, therefore, full of vectors and scalars, endowed with and permeated by them. Secondly, when to every point of the space occupied by a continuous flow there corresponds a vector $\bar{R}$, of definite direction and tensor, generally varying from point to point, the space is said to be a *vector field*, i.e. the seat of the vectors $\bar{R}$. When $\bar{R}$ stands for $\bar{v}$, we have a velocity field, i.e. the field of the velocity vector $\bar{v}$ representing at each point the direction and the absolute value of the flow. When $\bar{R}$ stands for $\bar{a}$, we have a field of accelerations. And so on.

A vector field is analytically continuous, if $\bar{R}$, both in value and direction, varies in a continuous manner from point to point and in time; in which case $\bar{R}$ – and its components (whatever they represent) – admit everywhere definite differential coefficients, at least of the first and second orders, with respect to space (co-ordinates), and time.

Having established these basic foundation stones, directly or by implication, Leonhard Euler embarked on the building of the edifice of mathematical operations itself. And here he displayed the prodigious ability of an absolutely towering master. As Lagrange put it, Euler 'did not contribute to Fluidmechanics but created it'. It is a matter of regret that the purpose of this book prevents us from reproducing here Euler's garden of roses of mathematical virtuosity. When I look at these brilliant differential equations, with their mirror-like partial derivatives, I cannot help feeling that Euler was for Fluidmechanics what Leonardo da Vinci or Rembrandt was for the arts.

It would, however, be unfair to our reader not to give here at least one or two examples of Euler's mathematical forms. They are:

the differential equation of continuity

$$\boxed{\frac{d\rho}{dt}+\rho\left(\frac{\partial v_x}{\partial x}+\frac{\partial v_y}{\partial y}+\frac{\partial v_z}{\partial z}\right)=0},$$

and the differential equations of fluid motion

$$\boxed{\begin{aligned}
\frac{dv_x}{dt}&=a_x=g_x-\frac{1}{\rho}\frac{\partial p}{\partial x}=\frac{\partial v_x}{\partial t}+v_x\frac{\partial v_x}{\partial x}+v_y\frac{\partial v_x}{\partial y}+v_z\frac{\partial v_x}{\partial z}\\
\frac{dv_y}{dt}&=a_y=g_y-\frac{1}{\rho}\frac{\partial p}{\partial y}=\frac{\partial v_y}{\partial t}+v_x\frac{\partial v_y}{\partial x}+v_y\frac{\partial v_y}{\partial y}+v_z\frac{\partial v_y}{\partial z}\\
\frac{dv_z}{dt}&=a_z=g_z-\frac{1}{\rho}\frac{\partial p}{\partial z}=\frac{\partial v_z}{\partial t}+v_x\frac{\partial v_z}{\partial x}+v_y\frac{\partial v_z}{\partial y}+v_z\frac{\partial v_z}{\partial z}
\end{aligned}}$$

These are the blood, the flesh and the bones of Fluidmechanics. It is remarkable that they have not changed at all since the day Euler derived them. By their discovery by Euler, wrote Lagrange, the whole mechanics of fluids was reduced to a matter of analysis alone, and if the equations ever prove to be integrable, the characteristics of the flow, and the behaviour of a fluid under the action of forces, will be determined for all circumstances.

Louis de Lagrange (1736-1813)

Lagrange's 'if' was overcome by Lagrange himself, and in a most masterly manner. Anyone who has ever tried to study his famous *Mechanique analytique* first published in France in 1788, will agree that he distinguished himself in the history of Fluidmechanics on such a scale that his name can easily be put side by side with Euler's; moreover, in some respects he surpassed Euler. Aristotle, Archimedes, Galileo, Torricelli, Stevinus, Pascal, Huyghens, Bernoulli, Clairaut, Descartes, d'Alembert, and many others contributed to the formation of the discipline; but Newton was the first to cement the foundation of the edifice, Euler to erect its walls and floor, and Lagrange to add all, or almost all, those other major parts which make an edifice a safe and rather enjoyable house. He became preoccupied with the organization and perfection of mechanics, of its mathematical language and methods.

'My objectives were', wrote Lagrange, 'to reduce the theory of mechanics and the art of solving the associated problems to general formulae and to unite the different principles in mechanics'. He succeeded in this with the glitter of a genius. However, I am mainly interested in his direct contributions to Fluidmechanics.

Euler considered the motion of individual fluid particles along their trajectories. But Lagrange felt that, since the number of such particles is infinitely large, the motion of each individual particle required to be specified in some way. To achieve this, he suggested choosing the initial (or starting) co-ordinates of a particle, at $t = 0$, as the characteristics of its motion.

Namely, let (a,b,c) be the co-ordinates of a fluid particle at $t = 0$. Then the trajectory of this particle (among the trajectories of an infinitely large number of particles) will be that one which will pass through point (a,b,c). Thus, the co-ordinates (x,y,z) of the point under consideration along its trajectory will be $x = x(a,b,c,t)$, $y = y(a,b,c,t)$ and $z = z(a,b,c,t)$, where (a,b,c,t) are known today as Lagrange's Variables. These equations represent a family of trajectories, which fill the whole region of flow, (a,b,c) being their parameters.

Thus, in the Euler method the velocity components of the fluid particle are functions of the co-ordinates (space) and time, while in the Lagrange method they appear to be given by the above equations. Mathematically speaking, $v_x = v_x(x,y,z,t)$, $v_y = v_y(x,y,z,t)$, and

$v_z = v_z(x,y,z,t)$ – in Euler's method, and $v_x = \partial x/\partial t$, $v_y = \partial y/\partial t$, and $v_z = \partial z/\partial t$ – in Lagrange's method. And the latter himself acknowledged that Euler's method was and remains more convenient and logical.

The trouble was, however, that the above three Euler differential equations of fluid motion contained five unknowns (v_x, v_y, v_z, p, ρ), therefore two additional equations were needed. Euler hinted and Lagrange showed that the differential equation of continuity and the equation of physical state could serve as such equations. So, according to the formal logic of mathematics, the equations must have been integrable. 'In the circumstances, Lagrange a man of great academic principles and pride, could not accept defeat, he had to develop a solution' (D. P. Riabouchinsky). And this is where we must return to our earlier question: who was the real author of the so-called Bernoulli Equation?

The notion 'total differential' was known already to Euler and to A. C. Clairaut (1713–65); the latter applied it to the solution of problems in fluidstatics.† But Lagrange was the first to develop it into a powerful tool of fluidmechanics. He came to the conclusion that Euler's equations could be solved only for two specific conditions: (1) for potential (irrotational) flows, and (2) for non-potential (rotational) but steady flows.

The first of these cases required the introduction of the so-called 'Velocity Potential' $\varphi = \varphi(x,y,z)$, such that $v_x = \partial\varphi/\partial x$, $v_y = \partial\varphi/\partial y$, $v_z = \partial\varphi/\partial z$. This was yet another revolutionary development in the formation of Fluidmechanics, which remains vital up to these days. The introduction of the velocity potential made it possible to carry out extremely interesting mathematical operations, and thus to reduce Euler's equations of motion to a single total differential equation,

$$d\left(\frac{v^2}{2} + \int \frac{dp}{\rho} + \frac{\partial\varphi}{\partial t} - d\Phi\right) = 0$$

The integral of this equation is

$$\boxed{\frac{v^2}{2} + \int \frac{dp}{\rho} + \frac{\partial\varphi}{\partial t} - d\Phi = C(t)}$$

This is, then, Lagrange's integral of Euler's equations of motion of an irrotational (potential) compressible fluid. The integral for the second case (steady flow) is similar. For a steady flow, $\partial\varphi/\partial t = 0$ and the time-dependent constant $C(t)$ becomes simply C. If, in addition, the flow is incompressible, $\int dp/\rho = p/\rho$. Thus,

† *Theorie de la figure de la Terre tiree des principes de l'hydrodynamique.* Durand, Paris, 1743.

$$\boxed{\frac{v^2}{2} + \frac{p}{\rho} - \Phi = C = \text{const}}, \text{ or } \boxed{\frac{\rho v^2}{2} + p - \rho\Phi = \text{const}}$$

Here $\Phi = \Phi(x,y,z)$ is the gravitational potential, such that $\partial\Phi/\partial x = g_x$, $\partial\Phi/\partial y = g_y$, $\partial\Phi/\partial z = g_z$, where g is the gravitational acceleration. If we choose the oy-axis vertically upward, then $g_x = g_z = 0$ and $\partial\Phi/\partial y = -g_y = -g$, therefore $\Phi = -gy$ and $-\rho\Phi = \rho gy$. Let us recall now the equation of state: $p = \rho RT$, or $\rho = p/RT$. Then $\rho gy = pgy/RT$, or $\rho gy/p = gy/RT$. Substituting here the well-known values ($T = 288°$ and $g = 9{\cdot}81$ m/sec^2), we find that the third term of the left-hand side of Lagrange's integral is negligibly small compared with the first and second terms, so that

$$\boxed{\frac{\rho v^2}{2} + p = \text{const}}$$

To sum up, we have shown that the so-called 'Bernoulli Equation', probably the most fundamental equation in the entire Fluidmechanics, ascribed to D. Bernoulli in every textbook known to us, is, in fact, not Bernoulli's at all, but LAGRANGE'S INTEGRAL OF EULER'S EQUATIONS OF MOTION.

In due course, however, other methods of development of the equation emerged.

Jean le Rond d'Alembert (1717-83)

Lagrange's integral of Euler's equations shows, as, indeed, was shown also by Daniel Bernoulli, that the greater the velocity at a given point of a flow, the less the pressure at that point, and vice versa. But, by Leonardo da Vinci's law, where the velocity is greater, the cross-sectional area is smaller, and vice versa.

Many interesting conclusions follow from these fluidmechanic laws. For example, our ancestors knew that rivers flow faster where they are narrower; but they were unaware of the fact that pressure p was lower in the narrower places. Then, experienced captains know that it is dangerous to keep two moving ships close to each other, because the cross-sectional area of water between them becomes small, narrow, therefore the water speed increases and, consequently, the water pressure decreases, which creates a danger of collision, of drawing them towards each other. Another example: almost every chimney has been built with a contraction at the exit; this contraction, this reduction of the cross-sectional area, results in an increase of the smoke velocity and, consequently, in a corresponding decrease of the pressure, which, in turn, creates a sucking effect, thus preventing smoke from escaping into the room.

Jean le Rond d'Alembert was the first to apply these phenomena to the study of the resistance offered by an ideal fluid to a body moving in it.† But the results of his almost heroic efforts were so complex mathematically, and so far above the grasp of the average reader, that they had little impact on the further development of Fluidmechanics. They produced, however, some valuable new ideas and theorems, which deserve attention.

The widely known and important notions 'Stagnation Point' (i.e. point on the surface of a body exposed to fluid flow where the flow velocity is zero) and 'Stagnation Region' were introduced by d'Alembert. This is how he described them: the particles moving along the central streamline towards *O* do not travel as far as *O*, they stop just before reaching the point, therefore at *O* and immediately in front of *O* the fluid is necessarily stagnant.‡

He was disturbed by his result, because he simply refused to understand 'the role' of a stagnant fluid, however small its quantity. In order to avoid this difficulty, he proposed a body configuration with an 'infinitely sharp leading edge' (Figure 45), so that 'there will no longer be

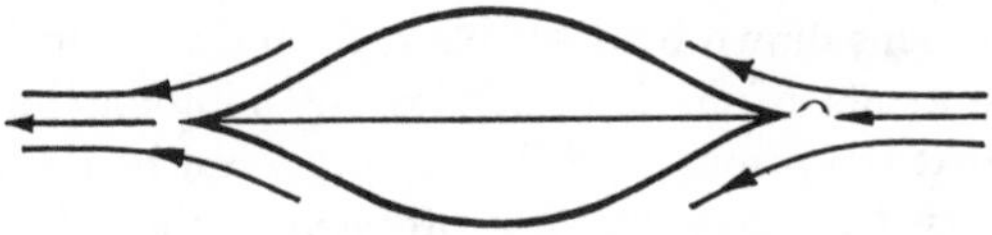

Fig. 45. d'Alembert's sharp-edged body configuration

stagnant fluid, and the whole fluid will run past the forward surface' without disturbance. Whether we should accept this as a prevision of the modern high-speed (sharp leading edge) aerodynamic configurations, is a matter of speculation and I leave it at that.

But what about the rear-side of the body? It may seem at first sight, writes d'Alembert, that the motion must be different here. But my theory of motion of fluids, he writes, shows that the velocity components in front and behind the body are exactly the same, therefore all the other conditions will also be the same. If so, he continues, providing mathematical proofs in his peculiar manner, 'the pressure of the fluid on the forward

† *Traite de l'Equilibre et du Mouvement des Fluides,* Paris: 1744.

‡ *Essai d'une nouvelle theorie de la resistance des fluides.* Paris: 1752.

surface is equal and opposite to the pressure on the backward surface, therefore the resultant pressure will be absolutely nothing'.

This is, then, the 'd'Alembert Paradox', which can be formulated more precisely as follows: the fluidmechanic resistance of a body moving steadily in an ideal fluid is zero. Its modern proof can be found in standard textbooks of Fluidmechanics, therefore we do not reproduce it.

D'Alembert was also the first to introduce something like the laminar flow concept, i.e. the concept of a flow composed of parallel slices of fluid. In general, let the velocities of the different slices of the fluid, at one and the same instant, be represented by the variable v. Then imagine that dv is the increment of the velocity in the next instant, the quantities of dv being different for the different slices, positive for some and negative for others. Or, briefly, imagine that $v \mp dv$ expresses the velocity of each layer when it takes the place of that which is immediately below. 'I say', wrote d'Alembert, 'that if each layer is supposed to tend to move with an infinitely small velocity $\pm\, dv$ (in relation to its neighbouring layers), the fluid remains in equilibrium.†

For since the velocity of each slide is supposed not to change in direction, each layer can be regarded, at the instant that v changes to $v \mp dv$, as if it had both the velocity $v \mp dv$ and $v \pm dv$. Now, since it only retains the first of these velocities, it follows that the velocity $\pm\, dv$ must be such that it does not affect the first and is reduced to nothing. Therefore, if each slide were actuated by the velocity $\pm\, dv$, the fluid would remain at rest.'‡

The conclusions drawn by d'Alembert from this analysis are of great interest to students of Fluidmechanics. In addition, he also proved analytically that the principle of 'living force' can be applied to fluids. He then analysed the now classical problem of the velocity of a fluid leaving a vessel which is kept filled to a constant height. And we should not forget, of course, the d'Alembert Principle, which states that Newton's third law holds for forces acting upon bodies entirely free to move as well as upon fixed bodies in stationary equilibrium.

Chevalier de Borda (1733-99) and others

We are moving closer and closer to modern Fluidmechanics, which emerged at long last, unfortunately, as a much more complex system

† *Essai d'une Nouvelle Théorie de la Résistance des Fluides.* Paris, David: 1752.
‡ *Traité de l'Equilibre et du Mouvement des Fluides.* Paris: 1744.

than has been described so far. As Chevalier de Borda, a French mathematician and nautical astronomer, remarked, real fluid flows are 'more sophisticated than the most sophisticated lady's character'. He sounded a warning, for example, that not all flows are 'in harmony' with Daniel Bernoulli's, Leonardo da Vinci's and Lagrange's laws, i.e. not always does an increase in the cross-sectional area lead to the proportional decrease in the flow velocity and increase in flow pressure. When (he says) a perfectly normal flow experiences a sudden expansion, which may happen in a pipe as well as on the surface of a body, it gets disturbed to such an extent that it loses a part of its kinetic energy, or living force'.†

This became known as the Borda Theorem (see Plate 1). Its implications for applied Fluidmechanics proved to be far reaching. For instance, the concept of *flow separation* became an organic part of the study of real flows, and it encouraged the emergence of a new branch of fluid flows – flows through valves and orifices of all kinds. It would be difficult to call them all Borda Flows, nor can we say that they were all studied during Borda's lifetime. But it may be relevant to describe them in connection with his contributions.

An orifice is an opening having a closed perimeter through which a fluid may discharge. It may be open to the atmosphere, which is the case of free discharge, or it may be partially or entirely submerged in the discharged fluid. Typical examples of orifices are shown in Figure 46 (*a*,*b*,*c*).

An orifice may be very small, as in the case of those used for leak

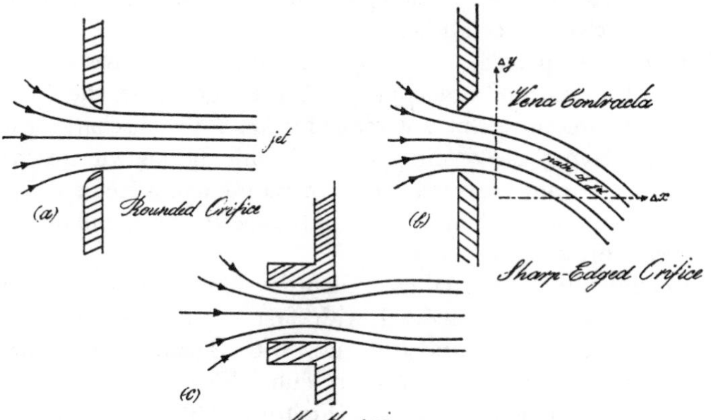

Fig. 46. Typical schemes of orifices

† *Mémoirs de l'Académie des Sciences*. Paris: 1766.

ports or for calibration, or large as illustrated by sluice gates in a dam. In all cases, the head of the fluid on the orifice is measured from the fluid level surface to the central line of the orifice.

The jet of a sharp-edged orifice converges from all directions, and so continues to converge for approximately one-half of the orifice diameter downstream. The contracted section is known as the *vena contracta*. The ratio of the cross-sectional area of the contracta to the area of the orifice is known as the contraction coefficient of the orifice.

Friction in the orifice slows the velocity to a somewhat lower value than the ideal free spouting velocity given by Torricelli's Law, $v = \sqrt{2gh}$. The ratio of actual to ideal velocity is the orifice velocity coefficient. Since the discharge is the product of velocity and area (Leonardo da Vinci's Law), the discharge coefficient is the product of the velocity and contraction coefficients. It has a numerical value of $\sim$61% for the average sharp-edged orifice.

The path taken by a jet discharging freely (horizontally) under the pressure of the head h is parabolic in shape due to the pull of gravity (Torricelli's Trajectory).

An orifice on the side wall of a tank near the bottom has a higher coefficient of contraction than one which is located further away from the bottom. Similarly, an orifice with the upstream edges rounded has a higher coefficient of contraction than one with sharp-edges. The discharge may be as much as 30% greater for well-rounded orifices. Orifices which are submerged, orifices which are squared instead of circular, and orifices in which the fluid approaches with a high velocity, cannot be treated by the equation given above without corrections being made for these special conditions.

Orifices are employed, for example, for measuring flows of vapours and gases. By means of a manometer, or pressure gauge, the upstream and downstream pressures are measured, and the discharge can be determined from those readings coupled with the known area of the orifice, the upstream pressure, the temperature, and a factor which involves gas constants, such as the ratio of the specific heats and upstream-downstream pressures.†

A contemporary of Borda, an Italian physicist called G. B. Venturi (1746–1822), is usually credited with first combining Leonardo da Vinci's and Bernoulli-Lagrange's laws with Torricelli's manometer into one device known as the 'Venturi Tube'. The tube has a first converging then diverging configuration (Figure 47). Its cross-sectional area

† *Van Nostrand Scientific Encyclopaedia*. Also, *Technische Strömungslehre*. von Bruno Eck, Springer-Verlag: 1966. *Hydraulics and its applications*. A. H. Gibson, Constable & Co. Ltd, London: 1930.

at A is large, therefore, by Leonardo da Vinci's law, the velocity here is small and, consequently, by the Bernoulli-Lagrange law, the pressure is high. The cross-sectional area at B is small, therefore the velocity is high here and the pressure low. Thus, the liquid in the U-shaped tube manometer is under high pressure at A and lower pressure at B, therefore the column in the A-leg is lower than in the B-leg.

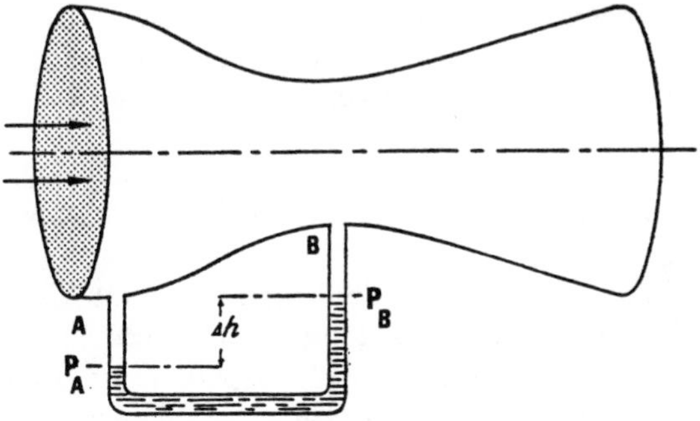

Fig. 47. The Venturi Tube, which is similar to the modern wind-tunnel

If we work out the mathematics of these phenomena, the following formula emerges:

$$\boxed{v_B = \sqrt{\frac{2\,(p_B - p_A)}{\rho\,(S_B^2/S_A^2 - 1)}}},$$

where S_A and S_B are the cross-sectional areas of the tube, p_A and p_B the pressures at the cross-sections, and ρ the mass density of the fluid. Thus, the Venturi Tube can, and does, serve as an instrument to measure the fluidflow velocity.

I should, perhaps, add that the post-Venturi development of this important device was carried out by Clemens Herschel (1842–1930), an American hydraulic engineer. Like Torricelli, Newton, Borda and others, Herschel concluded that no equation or measuring device will ever be reliable enough without taking into account the viscosity of the fluid studied.

Chezy, Du Buat, Coulon, Hagen, Poiseuille and Girard

The role of temperature in fluid friction – viscosity – was neglected by Newton, but studied subsequently by Du Buat, Girard and others, and especially by Poiseuille. They showed that the force of friction decreases with the increase of temperature.

It was Navier, 150 years after Newton, who first introduced viscosity into the general equations of motion (see below), and Poisson, both outstanding French mathematicians, who ascribed fluid friction to the change of the repulsive forces of fluid particles. Saint-Venant, Kleitz, Helmholtz, Meyer, Kirchhoff and others advanced another theory, according to which rapid displacements of particles in a flowing fluid create in it forces which are proportional to the relative velocities of the layers of the fluid.

According to Girard,† the first mathematical equation of a steady viscous fluid flow in a canal was written by a Frenchman, Chezy: $Qh = atv^2$, where Q denoted the cross-sectional area of the canal, h the gradient of the canal per unit length, t the wetted perimeter of the cross-section, v the mean velocity of flow, and a some constant coefficient. Coulon tried to compare this with Newton's fluid-resistance formulae $(av^2 + bv^{3/2} + cv)$ and $(av^2 + b)$, but failed to produce any reliable conclusions. Nevertheless, his extensive experiments proved to be valuable for some of his contemporaries and to many future investigators.

Denoting by d diameter, l length, and p pressure per square unit area of the cross-section, Girard proved mathematically that $(dp/4l) = av^2 + bv$. While checking this formula experimentally, he came to the conclusion that (1) the laws of water flow in long and short pipes are different; (2) in a sufficiently long pipe, p and v change so that the ratio $(dp/4lv)$ remains constant, $(dp/4lv) = b$ (this important result was confirmed also by Hagen‡ and Pouseuille,§ and, therefore, became an essential part of the famous Hagen-Poiseuille law, which will be discussed later on); (3) the minimum length l of a pipe needed to satisfy the law $(dp/4lv) = b =$ const increases with the increase of its diameter and pressure; (4) Chevalier Du Buat showed|| that the colder the water the

† *Mémoires de l'Institut de France*, p. 550, etc.: 1813.
‡ *Academie der Wissenschaften*, S.92. Berlin: 1854.
§ *Le Mouvement des Liquides dans des Tubes de Petits Diametres.* Paris: 1844.
|| *Principes d'hydraulique.* t. II, p. 9, edit.: 1816.

slower it flows; Girard's theory and experiments confirmed that of two flows satisfying the condition $(dp/4lv) = b$, the one with the higher temperature is faster; and (5) for flows in short pipes (not satisfying the condition) the influence of temperature can be neglected.

Girard's experiments disproved Coulon's theory that fluid resistance is proportional to v, and suggested the formula $(av^2 + bv)$, which, however, was not accurate enough for pipes of large diameters. All in all, Girard's investigations failed to produce results of lasting importance because his analysis of the 'lack of slipperiness' did not go far enough. And this was understandable: the discovery of the laws of motion of celestial mechanics offered Galileo much less difficulty than the study of a 'simple' water flow; giants like Euler and Lagrange developed beautiful theories for ideal fluids, but ignored the problem of viscosity. And Girard, Couplet, Chezy, Du Buat, Coulon, Proni, and many others, were not really of their stature.

But Jean L. M. Poiseuille (1799–1869) was a different figure. He did not realize, perhaps, that his thorough investigations of blood circulation and his *Le Mouvement des Liquids dans des Tubes de Petit Diametres* would put him among the prominent developers of Fluidmechanics, especially because, both by education and by profession, neither he himself nor his contemporaries, considered him anything but a physician.

'I began my investigations', wrote Poiseuille in the introduction,† 'because progress in physiology demands a knowledge of the laws of motion of the blood, which is to say, a knowledge of the laws of motion of fluids in small-diameter ($\sim$0·01 mm) pipes.' 'Of course', he continued, 'Du Buat, Girard, Navier and others have already studied this problem: but it is desirable to have further study and experimental investigation, so as to have reliable comparison of theory with experimental data.'

These are the numerical values of the limits of (l/d):

Diameters at one end, perpendicular to each other, mm	Diameters at the other end, perpendicular to each other, mm	lim(l/d)
0·0286 –0·0296	0·02933–0·0300	70
0·04466–0·0460	0·04250–0·0445	80
0·0845 –0·0845	0·0850 –0·0860	120
0·1117 –0·1135	0·1125 –0·1145	270
0·1395 –0·1415	0·1405 –0·1430	180
0·6160 –0·6932	0·6140 –0·6900	310

† *Poiseuille, Recherches expérimentales sur le mouvement des liquides, Memoires.* t. IX, p. 433, Inst. Acad. Royale des Sciences.

Girard's experiments showed, and now Poiseuille's experiments confirmed, that in pipes of diameters ranging from 0·0286 to 0·6900 mm, the character of a fluid flow depends on the value of (l/d), which is different for different diameters, and it decreases with the increase of d.

His second conclusion was that the amount of fluid Q passing in a pipe per unit time is proportional to pressure p, and may be expressed as

$$Q = 2495{\cdot}224\,\frac{pd^4}{l} \ldots \text{(at } t = 10°\text{C, and with } Q \text{ in mm}^3\text{)}$$

This was found also by H. Hagen† and Jacobson,‡ and therefore became subsequently known as the Hagen-Poiseuille formula, in the form

$$\boxed{Q = \frac{\pi r^4}{8\mu}\frac{dp}{dx}}$$

The Hagen-Poiseuille law gives reliably accurate results only for highly viscous flows, and in pipes of small diameter.

We shall return to these formulae a little later on. In the meantime, let us introduce another Frenchman whose role in the history and philosophy of fluidmechanics is comparable with that of Euler or Lagrange.

Claude Louis M. H. Navier (1785-1836)

The then rapidly growing fluidmechanics demanded, however, a more general solution of the problem of viscosity. Above all, it was necessary to establish the most general equations of motion of real, viscous, fluids. Yes, Euler was the creator of Hydrodynamics; yes, he revolutionized the science of fluid flows; but the beautiful trousers he tailored had no buttons, they failed to include viscosity.

The buttons were provided by Claude Navier, a great mathematician and analyst. In a paper read to the Academie des Sciences on 18th March, 1822,§ he explained that the equations and the solutions he put

† *Abhandl. der Kingl. Academie der Wissenschaften zu Berlin.* 2B, S.1 und 2. 1869.
‡ *Dr. Heinrich Jacobson Reicharts und Dubois*—Reymonds Archive, S. 99–100. 1860.
§ *Mémoirs de l'Académie des Sciences de l'Institut de France.* t. vi, Paris: 1822.

forward had necessarily to be very complicated. 'But although they are based upon Newton's hypothesis $\tau \propto (dv/dy)$', he said, 'it cannot be assumed that they represent nothing new.'

The mathematical techniques employed by Navier are very complicated, therefore I am omitting them completely. Instead, let us attempt to describe his basic ideas. Newton's 'lack of slipperiness' was due to molecular motion between the various portions of a flowing fluid. Namely, any pressure tends to reduce the distances between the molecules of the fluid. This creates intermolecular repulsive forces. If the fluid is at rest, the compressive and repulsive forces cancel out each other, and every molecule is in equilibrium. But when the fluid is in motion the intermolecular actions vary from one point to another, which is resisted by the fluid. In other words, in a fluid flow, the repulsive actions of the molecules are increased or decreased by an amount proportional to the velocity with which the distances decrease or increase.

Having thus devised a physical model, Navier then proceeded to the mathematical study of viscous flows, and developed the famous Navier Equations (which, in due course, were obtained also by other investigators, in different ways – Poisson, Stokes, Gromeka, and others).

Consider first a steady horizontal viscous flow in a straight pipe. Ducleuax showed experimentally that, in such a flow, (1) all fluid particles move along straight lines, (2) all fluid particles at equal distances (radii) from the centre at any cross-section move at equal velocities, (3) because of the 'lack of slipperiness', particles adjacent to the walls have zero velocities, and particles moving along the central line have the greatest velocities.†

These concepts enable the equation under discussion to be obtained in several ways, of which we choose the least difficult to understand, although, admittedly, very laborious. The second reason for this choice is that the method is based upon the same Eulerian infinitesimal parallelopiped.

Imagine, then, a particle of viscous fluid in the form of a prism (Figure 48). Its faces are now under the action not only of pressure, but also of viscosity, or friction; according to the Stevinus-Pascal law, pressures are normal to the faces, while the forces of friction are tangential to them. Thus, the resultant forces acting upon the faces will be inclined to the faces.

Let $\bar{p}$ be the resultant stress vector on a face; we can then denote its components by p_{xx}, p_{xy}, p_{xz}, etc., each of which represents a stress on the plane perpendicular to the axis of co-ordinates symbolized by the first

† *Ecoulement de divers liquide au travers des espaces capillaires.* Annales de chimie et de physique, t. xxv, 1872, Paris.

suffix, in the direction of the second suffix. So that, for example, p_{xy} is a stress in the plane perpendicular to *ox*, in the direction *oy*, and is therefore a tangential stress on this plane.

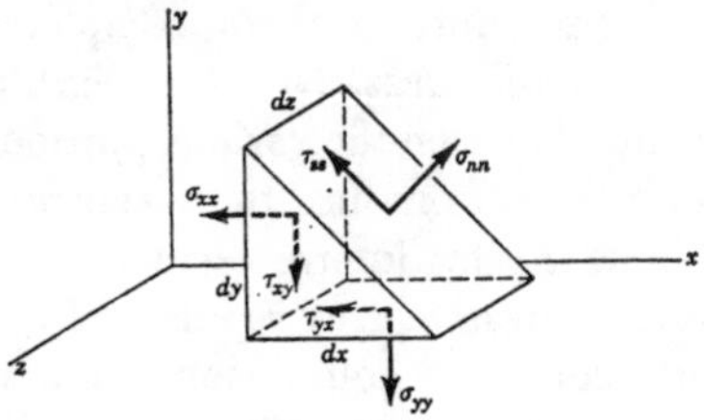

Fig. 48. Navier's infinitesimal viscous fluid particle

By this notation, the stresses p_{xx}, p_{yy} and p_{zz} are normal to the corresponding faces of the parallelopiped, whereas p_{xy}, p_{xz}, p_{yx}, p_{zy} are tangential stresses. But we have already seen that $p_1 = p_2 = p_3$ (Stevinus-Pascal law); in our present notations this means that $p_{xx} = p_{yy} = p_{zz} = -p$, a simple but important fact, which made Navier's task easier.

Having similarly found the stress components for the other faces, he then applied to the remaining forces Newton's second law and, after very laborious mathematical operations, arrived at the following historic differential equations of real (viscous) fluid flows:

$$\frac{\partial v_x}{\partial t} + v_x\frac{\partial v_x}{\partial x} + v_y\frac{\partial v_x}{\partial y} + v_z\frac{\partial v_x}{\partial z} = g_x - \frac{1}{\rho}\frac{\partial p}{\partial x} + \nu\left(\frac{\partial^2 v_x}{\partial x^2} + \frac{\partial^2 v_x}{\partial y^2} + \frac{\partial^2 v_x}{\partial z^2}\right)$$

$$\frac{\partial v_y}{\partial t} + v_x\frac{\partial v_y}{\partial x} + v_y\frac{\partial v_y}{\partial y} + v_z\frac{\partial v_y}{\partial z} = g_y - \frac{1}{\rho}\frac{\partial p}{\partial y} + \nu\left(\frac{\partial^2 v_y}{\partial x^2} + \frac{\partial^2 v_y}{\partial y^2} + \frac{\partial^2 v_y}{\partial z^2}\right)$$

$$\frac{\partial v_z}{\partial t} + v_x\frac{\partial v_z}{\partial x} + v_y\frac{\partial v_z}{\partial y} + v_z\frac{\partial v_z}{\partial z} = g_z - \frac{1}{\rho}\frac{\partial p}{\partial z} + \nu\left(\frac{\partial^2 v_z}{\partial x^2} + \frac{\partial^2 v_z}{\partial y^2} + \frac{\partial^2 v_z}{\partial z^2}\right)$$

In a different form, the same equations were also obtained by Sir George Gabriel Stokes (1819–1903),† a British mathematician and physicist, therefore they are often called the Navier-Stokes Equations. The basic mathematical philosophy of Fluidmechanics was thus complete; but not its history, because there were still many missing links.

† *On the theory of the internal friction of fluids in motion.* G. Stokes, Transactions of the Cambridge Philosophical Society, 8. 1845.

The birth of experimental fluidmechanics

Indeed, man had still to learn how to solve the Navier-Stokes equations, how to determine the fluidmechanic resistance of bodies moving in real fluids, how to deal with compressibility effects, etc., etc. D'Alembert bitterly complained that he, an accepted mathematician, was unable to say how to apply 'certain celebrated equations' to the determination of the air resistance of a simple body. It was, perhaps, for these and similar reasons that the Academie des Sciences of Paris created in 1775 a special Learned Committee, whose task was 'to investigate the possibility of improving the navigational qualities of ships'. The composition of the Committee was: d'Alembert (Chairman), Antoine Condorcet (1743–94), Abbé Charles Bossut (1730–1814), Adrian Marie Legendre (1752–1833) and Gaspard Monge (1746–1818).

Goethe once said that wisdom is only found in truth; and truth never grows old. D'Alembert's Learned Committee displayed precisely these two qualities, therefore it deserves a prominent position in the history and philosophy of Fluidmechanics. Its report† advised the builders of the subject that neither theories alone nor experiments alone will ever solve the complex problems of fluid flows; that the only right course was the combination of the two; and that in order to do this, experimental techniques should be developed up to the level of the theoretical achievements.

'The whole business of experimental investigations is so delicate a matter, however,' the Report warned, 'that it requires a very special attention, while in actual fact it often appears to be accidental. Information and data collected by many superficial investigators often appear to be unrelated to each other. In many cases, it is difficult to understand the causes and the sources of the data. Some forget the purpose of science. . . . It is necessary to be careful with the information presented by an experimentor who lacks theoretical principles; such an experimentor lacks vision and reasoning, and therefore often presents one and the same fact in different guises, without realizing it himself. Or he gathers at random several facts and presents them as proofs, without being able to explain them. It must be understood that scientific knowledge without reasoning – that is, without theory – does not exist.'

What a brilliant philosophy! What an introduction to Experimental

† *Nouvelles expériences sur la résistance des fluides*. Jombert, Paris: 1777.

Fluidmechanics! And what advice to give to those young and sometimes not-so-young scientists who rush to experiment before they realize what they are about!

Bossut, whose name has appeared on the preceding pages, was, perhaps, among the first to whom the word 'investigation' meant a definite logical sequence: *first theory, then experiment, and never vice-versa.* For it is theory, he said, which indicates what experiment is needed. Accordingly, he first worked out his theoretical results, and then proceeded to tow ship models and other bodies in the water basin of the Paris military academy (100 × 53 × 6½ feet).

He knew, of course, that a model submerged only partly experiences both liquid and air resistance. He also took into account the friction of the towing devices. To eliminate the air resistance (which he called 'the impact of the air'), he measured that part of the model surface which was not submerged, and assumed that the impacts of the water and air were, respectively, in 'compound proportion' to the impacted surfaces and the densities of the fluids.

Bossut was also aware of the role played by viscosity in fluidmechanic resistance. 'Friction between a flowing fluid and body surfaces takes place along the whole body', he wrote. 'Therefore, if we could get rid of friction, the slightest force would set the body in motion.' He was, however, wrong in his deduction that the 'stickiness' of water is 'extremely small': we shall see in due course that the opposite is the truth.

A military engineer named Du Buat (1734–1809), a contemporary of Bossut, published in 1779 a book called *Principes d'hydraulique*, which analysed the flow of water in rivers, canals and pipes, and studied the role of viscosity rather fully and systematically. According to him, flow resistance was due, to a large degree, to friction between the fluid and walls, causing 'retardation of the fluid'. This retardation is communicated from the wall to those parts of the flow which are not directly in contact with it; and it is this which affects the value of the mean velocity of the flow and determines the velocities near the wall.

It may thus be reasonable to say that Du Buat was yet another 'first father' of the boundary layer theory. On the basis of comparison of theory with experiments, he produced the theorem that the friction of a fluid is independent of its pressure. And it was, by the way, this important theorem which led him to another, but totally wrong, theorem: that the fluidmechanic resistance of walls (and surfaces generally) does not depend on the surface roughness.

Du Buat loved experiments. He studied flows in small canals and glass pipes of different sizes and configurations. His measuring techniques were poor, but his qualitative observations were effective. We know, for example, that he established an important law that the fluid-mechanic resistance of the walls was proportional to v in powers smaller than square, and gave a formula whose accuracy was surpassed only by the formulae of Darcy (1857) and Bazin (1869).

When you analyse the work of the old Mesopotamian (Assyrian) or Caucasian (Armenian) watermill (Figure 49), or of the Marcus Pollio

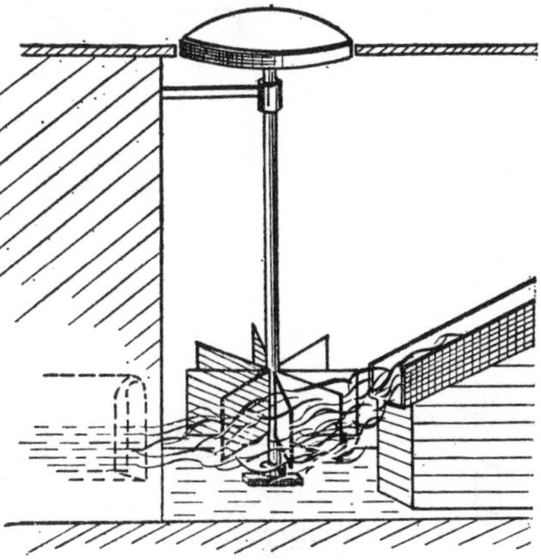

Fig. 49. The Caucasian-Persian water-mill

Vitruvius' watermill, which in due course became known also as the Leonardo da Vinci Wheel (Figure 14), or the old Persian water-lifting machine (Figure 50), you notice one common phenomenon; a number of the vanes, paddles and water-lifting caps are permanently immersed in water, fully or partly, and move in it. Arnold Sommerfeld (1868–1951) once said that this could be the explanation of the interest in the fluid-mechanic resistance experienced by flat plates moving in water. Perhaps it was, but how can we be sure?! Whatever the reasons, the historic fact is that flat plates were the first bodies studied experimentally. Already in the seventeenth century Edme Mariotte (1620–84) had measured the force acting on such a plate exposed to a stream of water, and Borda moved flat plates of various dimensions in stationary water by means of a rotating arm (Figure 51). But then Borda and Bossut, probably independently of each other, found themselves face to face with this question: can one and the same body have the same resistance, or drag, in finite and infinite conditions? Both gave the same answer: no. To prove the point, Borda carried out a whole series of experiments with models of equal or equivalent 'middle areas', i.e. maximum maximorum areas perpendicular to the direction of motion.

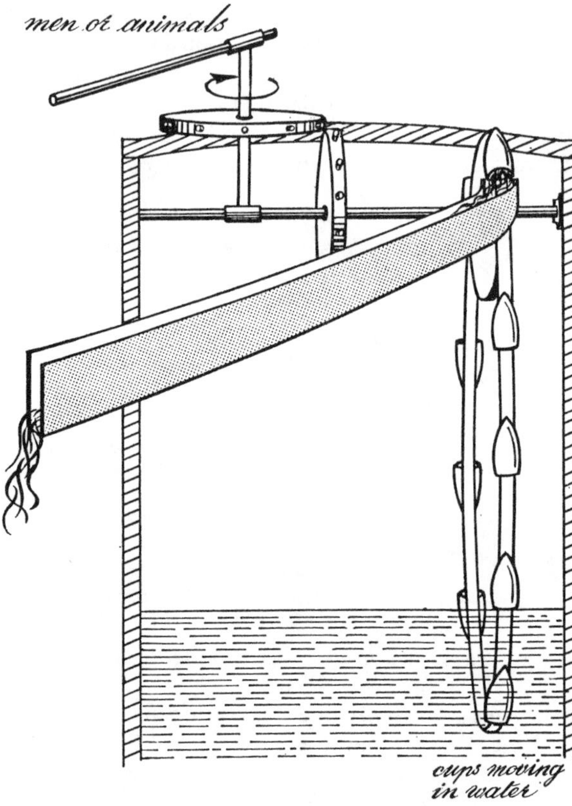

Fig. 50. The old Persian water-lifting machine

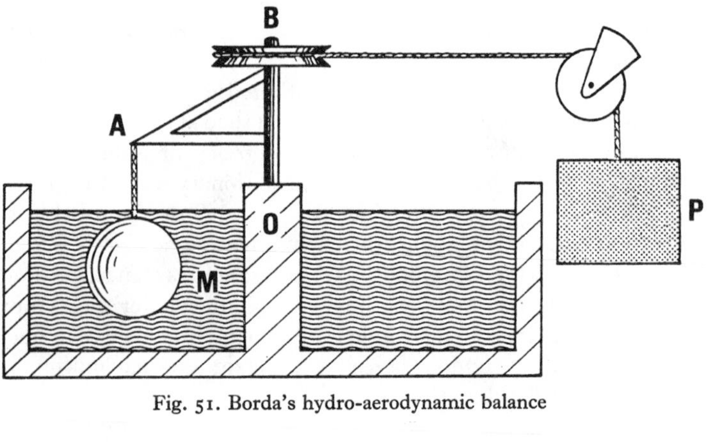

Fig. 51. Borda's hydro-aerodynamic balance

Figure 51 shows the hydro- and aerodynamic balance used by Borda. Under the pull of the weight *P*, wheel *B* spins and drives the arm *AB*, which tows model *M* immersed in water in a circular basin. The speed of one and the same model could be varied by means of various weights *P*. Models could be immersed fully (to study the water resistance alone), or partly (so as to study the combined water-air resistance), or not immersed at all (so as to study the air resistance alone).

I should, perhaps, say at once that, as the reader will see later on, this method of experimental investigation of the law of air resistance was used also by an Englishman called Robins. His machine, a kind of aerodynamic balance, is shown in Figure 52. Borda is often called 'the

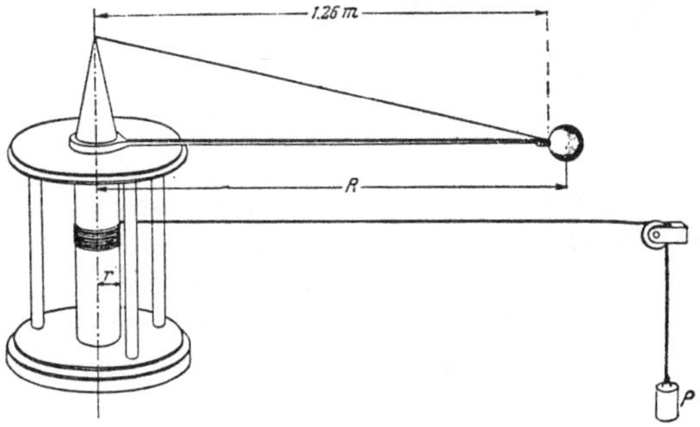

Fig. 52. Benjamin Robins' aerodynamic balance

father of experimental aerodynamics', but I am not sure at all whether it would be less accurate to associate this honour with Robins. Indeed, although Borda carried out a somewhat extensive programme of tests,† and although his pioneering contributions proved to be of great importance, it must be confessed that Robin's role in the formation of experimental aerodynamics were somewhat more impressive, more concentrated, more conclusive.

Before Borda's experiments, it was thought that the net drag of a combination of bodies could be computed by mere addition of the drags of constituent bodies. Borda was the first to point out that this was not so. The total drag of two spheres put close to each other and in that state moving in water or in the air is generally different from the sum of the drags of the two bodies when tested separately, he said. Today we know this phenomenon as fluidmechanic interference. It arises from the change in flow pattern which accompanies the placing of two bodies in close proximity. The interference, or the *Borda Effect*, may be favourable or unfavourable. A typical example of the last case is given in

† *Mémoires de l'Académie des Sciences*. Paris: 1763.

Figure 53: when the two well streamlined bodies move in a fluid separately, they experience almost no flow separation and consequently are almost free of the so-called Karman Vortex Street (see page 168). But when they are put together, the vortex formation behind the composite configuration is obvious.

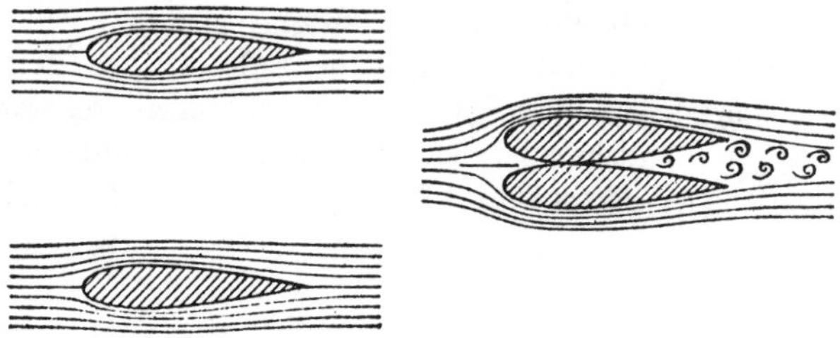

Fig. 53. Illustration of Borda's effect

Consider now an aircraft wing. At a large angle of attack, the flow separation on the upper side is strong and the vortex street distinct (see Plate 2). If, however, we attach to that wing one or two of the many auxiliary devices, which will be discussed later, the flow pattern over the upper surface improves significantly and the interference becomes favourable.

Aerodynamicists know how complex and important the interference problem is, especially in aircraft design. The wings interfere with the flow pattern of the tail-plane; the fuselage interferes with the flow patterns of the wings and the tailplane; and so on.

The other original contribution to Fluidmechanics made by Borda was his theorem that aerodynamic resistance was proportional to the square of the velocity v, as in Newton's formula, and to the sine of the angle of attack a, not as in Newton's formula ($R \propto v^2 \sin a$). This is known as *Borda's Law*.

Benjamin Robins (1707-51) and Leonhard Euler

In 1742, in London, Benjamin Robins published a book on the *New principles of gunnery containing the determination of the force of gunpowder and investigation of the difference in the resisting power of the air to swift and slow motions*, which, I think, was (at least in parts) the first printed work on what became in due course known as Experimental Aerodynamics. To reinforce this statement, I should like to invite Leonhard Euler to the witness box. Robins' book impressed Euler so much that he not only translated it into German, but enriched it with his own comments, corrections and important additions. In fact, his additions made the book five times larger (720 pages) and it was published under a new name.† 'I am a theoretician', Euler wrote in the Introduction, 'but I am sure that theoretical results should be endorsed by experimental proofs. Unfortunately, however, circumstances prevent me from dedicating the necessary time to experimental investigations. But Mr. Robins' experiments with the use of the ballistic pendulum were so numerous, and the results obtained by him are so important, that, I think, they will meet my needs, especially in relation to the study of the problem of motion of gun projectiles under the decelerating action of air resistance.'

Now, what were his interests? I have in front of me a collection of his papers, that show unmistakably that Euler probably studied air resistance before the appearance of Robins' book, perhaps in 1727, in the St. Peterburg Academy of Sciences (Russia). Here, in this Academy, Daniel Bernoulli tried to study the law, or laws, of vertical motion (upward and downward) of spherical bodies. Euler, his closest friend, elaborated Bernoulli's results mathematically, and used the v^2 law for the determination of air resistance. This particular work of his was published much later,‡ and therefore remained unknown to researchers like Du Buat, Bossut, Darcy, Bazin, Navier, Poiseuille, Borda, and others. The same v^2 formula was used by him in yet another work,§ which stated that as long as the medium (air) remained uniform and unaffected physically by the projectile itself, the square velocity law of air resistance was valid.

But let us return to Robins' *New Principles of Gunnery*. It was an outstanding work. Its second half was dedicated to experimental exterior ballistics. It, too, used the $R \propto v^2$ law and, still more interesting, came out with what is known today as the *Magnus Effect*. Gustav Heinrich Magnus (1802–70), a German chemist and physicist about whom more will be said later on in connection with the Flettner

† *Neue Grundsatze der Artillerie enthaltend die Bestimmung der Gewald der Pulvers nebst einer Untersuchung uberden Unterscheid des Widerstands der Luft in schnellen und langsamen Bewegungen, aus dem Englischen des Hrn. Benjamin Robins ubersatzt und mit den notigen Erlauterungen und mit vielen Anmerkungen versehen.* L. Euler, Berlin: 1745.

‡ *Meditatio in experimenta explosione tormentorum nuper instituta.* L. Euler, St Peterburg: 1862.

§ *Mechanik oder analytische Darstellung der Wissenschaft von der Bewegung mit Anmerkungen und Erlauterungen.* L. Euler, Berlin, 1748.

rotorship, was asked by the Prussian artillery authorities to investigate the reasons for the deviation of projectiles from their prescribed trajectories; he came out with the conclusion that a projectile rotating in flight creates for itself a sideways thrust, which compels it (the projectile) to deviate in the direction of that force; and this is presented by authors of books and articles on Fluidmechanics as the Magnus Effect. But they may be forgetting that the task of the philosophy of science is to make clear the content of scientific propositions, that is, to determine or discover their true meaning. The final determination of the meaning of propositions cannot be made by assertions, cannot itself form a science, because we would then have to ask ever anew for the meaning of the assertions, and would thus arrive at an infinite regress. Every statement of a meaning must, generally through scientific facts, lead to provable propositions.

What are, then, the facts? They are available for anyone's examination, and show beyond doubt that the Magnus Effect was discovered, and studied, and properly reported, by Benjamin Robins, some 110 years before Gustav Magnus.† As can be seen from Figure 54, the effect,

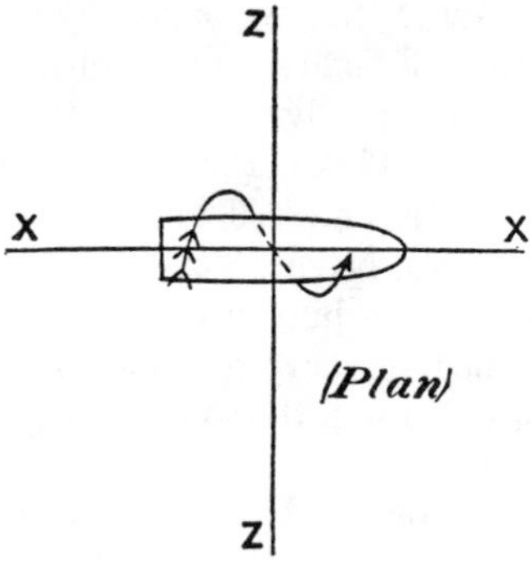

Fig. 54. Lanchester's scheme of flight of a bullet

the deviation of a projectile, is due, according to Robins, to the fact that: first, the projectile encounters aerodynamic resistance to its forward motion; second, it rotates about its longitudinal axis, thereby creating a degree of dissymmetry of flow. This analysis appeared in print in 1742, and was reported by Robins to The Royal Society in 1749.

But, how and why could Robins, an artillerist, arrive at this purely aerodynamic conclusion? The answer is that he advanced a theory of

† *Traité de mathématiques*. B. Robins, Traduit de l'anglais par M. Dupuy fils, Grenoble: 1771.

spirally-grooved barrels, and, consequently, spinning projectiles. Leonhard Euler, on the other hand, was against such a theory, therefore an argument developed between them. Robins was still relatively unknown: Euler was a towering figure. For this reason, Euler's point of view prevailed. Robins' was rejected: a tragedy not rare in the history of science and technology.

And what did Euler advocate? He insisted that all the deviations of projectiles were due to exterior form irregularities. This was, of course, a different philosophy. But it could not then, and cannot today, be dismissed as baseless. You may say that the interior surface of a barrel is made with a high degree of precision, argued Euler. You may also say that the exterior surface of a gun shell or of a rifle bullet is almost perfect. 'To which I reply that no man-made surface can ever be absolutely perfect: therefore shell-projectiles and bullet-projectiles can acquire rotational motion.'

Thus, it would, I think, historically be more accurate to talk about the 'Robins' effect' or, perhaps, about the 'Robins-Euler effect', than about the 'Magnus Effect'. But I am not sure that this recommendation will prove to be strong enough to overcome the inertia of tradition.

Robins established a number of experimental facts. His data proved, for example, that at high speeds (of rifle bullets, for instance) the air resistance increased faster than by the $R \propto v^2$ law, and he gave an explanation worthy of modern gasdynamics. There were two kinds of fluids, he said: first, those fluids whose densities are so high, or which are so compressed, that their particles manage to follow the fast-moving solid body and to fill immediately the vacuum created next to its base; second, those fluids which are highly rarefied and whose particles are therefore not connected with each other, they are unable to follow the fast-moving body and to fill the vacuum formed behind its base. In the last case, according to Robins, the air resistance must be greater than in the first.

This conclusion may seem to be somewhat naïve. But let us bear in mind that we are dealing with history which requires no approval or disapproval. We can, however, and we should look into the deep source of the conclusion. You see, Robins' physical model of the atmosphere was basically different from Newton's model. The air, according to Robins, consisted of infinitesimal particles in a state of constant motion. He considered the air as a rarefied medium, with large distances between its particles, which was, therefore, capable of being compressed by a fast-moving projectile. And when it becomes compressed, he said, for instance, immediately before the head of a projectile, the latter finds itself face-to-face with a totally new force, the force of elasticity of the air. . . .

Leonhard Euler approaches the same problem, once again, from a different standpoint. To him, a solid body (projectile) moving in the air decelerates (loses its speed) because 'it displaces the particles of the air'; the greater the number of the displaced fluid particles, the greater the loss of velocity, and vice versa. Hence, his general philosophical concept: he who wants to know the laws of motion of fluids, of solid bodies in fluids, and of fluids past solid bodies, should study the laws of motion of fluid particles.

The first problem Euler solved coincided with that considered by Newton.† It was the problem of motion of bodies in a certain rarefied medium. Like Newton, Euler considered this medium as consisting of a large number of infinitesimal particle-molecules uniformly distributed in space and being in a state of static equilibrium. The molecules were not connected with each other, and did not influence each other. He realized, of course, that there were in nature no such fluids; but he wanted a model for the development of a mathematical law of air resistance. And here are the two laws he derived:

$$\boxed{R_C = CS\rho v^2} \quad \ldots\ldots \text{ for a cylinder, and}$$

$$\boxed{R_P = CS\rho v^2 \sin^2 \alpha} \quad \ldots\ldots \text{ for an inclined flat plate,}$$

where C is a dimensionless coefficient of proportionality, S characteristic area, ρ mass density, v velocity.

Then, like Robins, Euler states: as velocity v increases, fluid particles are being compressed more and more; at a certain v, the distances between them diminish, and immediately in front of the body fluid pressure becomes high, while behind the body the opposite is the case. So, the air resistance of the body is higher at high speeds, and lower at low speeds.

From these very similar conclusions (endorsed by Robins' experiments), both Robins and Euler derived yet another conclusion widely used in modern gasdynamics. Namely, in order to prevent this sharp increase in the air resistance, it is necessary to reduce the amount of air compressed before the solid body; this can be achieved by means of reducing the frontal area of the head of the body, i.e. to use bodies with pointed and/or sharp edges.

We shall see later on that, indeed, when a body moving in the air reaches a certain speed, the air before it becomes compressed and the

† *Ballisticheskiye Issledovaniya Leonard Eilera.* A. P. Mandryka, Moscow: 1958 (in Russian).

so-called 'wave drag' causes a sharp increase in the air resistance. Robins did not use these words, the notion gasdynamics was still unknown, but the implications of his theory, definitions and experimental facts make it possible to say today that modern gasdynamics begins with him – and with Euler.

Lazare Carnot (1753-1823), Pierre Simon de Laplace (1749-1827) and others

We thus see that Euler in Russia and Robins in England knocked at the door of compressible aerodynamics quite firmly and unmistakably. Furthermore, Euler wrote† that one of the fundamental tasks of fluid-mechanic theory was to establish a definite mathematical relationship between the density ρ, elasticity ϵ, and pressure p, at each point, under the action of the 'accelerative forces'. And he did develop such a relationship:

$$dp = Ldx + Mdy + Ndz,$$

where $p = p(x,y,z)$ – pressure, $L = \partial p/\partial x$, $M = \partial p/\partial y$, $N = \partial p/\partial z$. If $p = p(x,y,z)$ is a given function of ρ at each point, the above equation becomes

$$\boxed{dp = \rho(Pdx + Qdy + Rdz)}$$

where P, Q, R are the rectangular components of the 'accelerative force'. These equations must be integrable when the density ρ is constant or uniquely dependent on p, he concluded.

We have already shown one method of such an integration (see Lagrange's Integral of Euler's Equations of Motion, page 79). The second method is as follows. If we assume that the air is in an adiabatic state, the relationship between its density and pressure is given by $p = \text{const} \times \rho^\gamma = C.\rho^\gamma$, or $\rho = \rho_0(p/p_0)^{1/\gamma}$, where p_0 and ρ_0 are the known pressure and density in the undisturbed air, and γ is the ratio of the specific heats of the air. Substituting this expression in the second term of the Lagrange Integral, and carrying out additional trivial mathematical transformations, the following fundamental equation emerges:

† *Principes généraux de l'état d'équilibre des fluides.* Mem. de l'Académie de Berlin: 1755.

$$\frac{p}{p_0} = \left(\frac{1 + \frac{\gamma - 1}{2} M_0^2}{1 + \frac{\gamma - 1}{2} M^2} \right)^{\gamma/(\gamma-1)},$$

where $M_0 = v_0/a_0$ and $M = v/a$ are the corresponding Mach numbers, which will be discussed later on.

Although Daniel Bernoulli had nothing to do with it, textbooks and journals call it again 'the Bernoulli Equation' for compressible flows. But if it is necessary to name the equation after someone, I would call it 'the Euler-Lagrange Equation'. The more important point is, however, that the ratio $\rho_0/\gamma p_0$ in this equation is equal to $1/a^2{}_0$, or $\gamma p_0/\rho_0 = a^2{}_0$, a_0 being the speed of sound. In other words, compressible fluid flows appear to be associated with the speed of propagation of sound waves in them (we shall return to this problem later).

Laplace was a French mathematician and astronomer. Among many other contributions to knowledge, he studied the theory of motion of the moon, the dynamics and fluidmechanics of tides, the theories of the specific heats, equilibrium of a rotating fluid mass, etc. Carnot, also known during his lifetime as *le grand Carnot*, was of a different kind altogether: a statesman and a general, deputy to the Legislative Assembly (1791) and National Convention (1792), member of the Committee of Public Safety (1793) and Commander-in-Chief of the Armed Forces (1793–5), and Minister of War (1800–1), Napoleon's Minister of Interior (1815) and so on. In 1815 he was exiled by Louis XVIII and lived first in Poland and then in Switzerland. Could such a busy man also be a mathematician? Yes, of course: Carnot's *Essai sur les machines en general*, *Principes generaux de l'equilibre et du mouvement* (1803), and *Mecanique analytique* (1788) constituted valuable pages in the history and philosophy of Mathematics, Mechanics and Fluidmechanics. It is interesting to mention that he was one of the few who dared to challenge Euler, Lagrange, d'Alembert and other recognized figures.

He thought that a scientist must also be a philosopher, and a philosopher also a scientist. This developed in his mind prejudiced attitudes towards many pioneers of knowledge. He 'accepted' Leonardo da Vinci as a great man, perhaps even as the greatest man, but in the philosophical sense only, because 'the confusion of the notion "force" could enter a mind that was only philosophical'. While Aristotle was to Carnot a 'mere thinker rather than a scientist'. And what about Sir Issac Newton himself? Did he not make a salad of the same notion 'force'? Is it not a fact that according to Newton, bodies attract each other without having anything in the space between them?

Yes, they all – Aristotle, da Vinci, Newton, Euler, d'Alembert and the rest – proved to be too small to understand the great thing called force. While he, Lazare Carnot, 'always knew' that nothing could be clearer than force. For (he said), force is always associated with motion, and the latter is as natural as nature itself. But when I say 'motion' (he continued), I have in mind its *quantity* and *direction*. The first of these can be measured only as the product of the mass of the body in motion and of the velocity with which it moves, mv; the second is determined by the velocity, which should, therefore, be denoted as $\vec{v}$, the arrow being a reminder that the movement has a direction. Thus, the quantity of motion is $m\vec{v}$, which should be called momentum . . ., and whenever and wherever there is momentum, there is also force . . . the two being associated by the formula

$$\boxed{\vec{F}\,dt = d(m\vec{v})}$$

Apart from the arrows, this is, in fact, Newton's formula.

But it was not merely given by Carnot: he derived and proved it. This is not surprising at all for Carnot was one of the few people of his time to whom a theory divorced from practice was like a Queen deprived of her kingdom. We can trace this state of mind and his flow of thought both in his writings and engineering-technological ideas. His *Reflections sur la Puissance motrice du feu* introduced us to thermo-dynamics, which is today the root of gasdynamics. Anyone familiar with this branch of knowledge knows about the famous Carnot Cycle, an ideal cycle of four reversible changes in the physical condition of the combustion substance: (1) an isothermal expansion, at constant temperature; (2) an adiabatic expansion, a process, without loss or gain of heat; (3) an isothermal compression; and (4) an adiabatic compression. Subsequent history showed that this was not only the *B*, if not the *A*, of the theory of internal combustion engines, but also an outstanding case of the philosophy of energy. It also put man face-to-face with the following interesting concept: no internal combustion engine operating between two given temperatures can be more efficient than a perfectly reversible engine operating between the same temperatures (Carnot's Theorem).

The detailed analysis of these concepts, and the basic considerations from which they were derived, indicate that Carnot was, indeed, more than a mathematician, a physicist and an engineer: he was a philosopher of the kind that was still unknown and rare, a philosopher of science, i.e. a man capable of deriving the most general concept based on actual knowledge and, consequently, capable of leading to practical propositions, to engineering creations. It is, therefore, both inaccurate and unfair historically to depict him as one who 'cared very little about the theoretical side of his problem' and as one whose ideas 'had no great influence on the progress of science'.†

The Carnot Cycle in itself was, and remains, an outstanding milestone in the history and philosophy of science and technology. But he also showed that the motive power of water depends on the quantity of water and the height from which it falls. To mention just one example, the water-mills which we have already discussed, provided adequate experimental proof. Yes, he was wrong in comparing the motive power of heat with that of water; but could any historian name a single scientist of any epoch who did not make mistakes? Then, was he not one of the builders of modern geometry? Could anyone say that thermodynamics had no great influence on science and technology? And what about his impulse-momentum equation: what branch of modern science can do without it?

And now about Laplace and Poisson. Simeon Denis Poisson (1781–1840), another great French applied mathematician, author of fundamental works on definite integrals, Fourier series and calculus of variations, without which Fluidmechanics and the theory of rocket trajectories would not be what they are, made the mathematical techniques of many branches of knowledge attractive and powerful. Nevertheless, I intend to restrict my references to him to his partial differential equation $\nabla^2\varphi = -u$, or

$$\frac{\partial^2\varphi}{\partial x^2} + \frac{\partial^2\varphi}{\partial y^2} + \frac{\partial^2\varphi}{\partial z^2} = -u$$

Poisson and Laplace were almost of the same age. In a way, and in a degree, they were also each other's pupils. I have in mind, for instance, the well-known Laplace's partial differential equation, $\nabla^2\varphi = 0$, or

$$\frac{\partial^2\varphi}{\partial x^2} + \frac{\partial^2\varphi}{\partial y^2} + \frac{\partial^2\varphi}{\partial z^2} = 0$$

Comparing the two equations, we see that when $-u = 0$, Poisson's equation becomes Laplace's equation; when $-u = 0$, Laplace's equation becomes Poisson's equation. Whichever of the two equations one takes, but especially the last one, it proves to be of fundamental importance. The flow past a body shaped for low resistance comprises two dissimilar parts: (1) a thin boundary layer, dominated by viscosity and

† *The Growth of Physical Science*, p. 268. Sir James Jeans, Cambridge: 1947.

merging into the wake, and (2) an external motion, in which viscous effects are scarcely measurable.

In the external – undisturbed – flow, the fluid is usually assumed to be devoid of viscosity, so that at any point the pressure acts equally in all directions; and if such a flow is irrotational, it will remain irrotational in flowing past an immersed body. Thus, the total pressure head given by the Lagrange Integral remains constant throughout the flow. To take account of the shape of the body, it is necessary to suppose that its surface is closely approached, but not so closely as to enter the boundary layer. This is tantamount to assuming that the boundary layer is everywhere very thin and that no wake exists. In the limit the fluid may be regarded as slipping with perfect ease over the surface of the immersed body. The boundary condition for this idealized fluid is, then, simply that the velocity component normal to the surface vanishes, and the whole flow is potential.

The mathematical theories of fluid flows of Clairaut, Stevinus, Euler, Lagrange and d'Alembert were based on this particular class of flows. Study of streamline flows is greatly facilitated by the velocity potential $\varphi(x,y,z)$, which gives the velocity components as $v_x = \partial\varphi/\partial x$, $v_y = \partial\varphi/\partial y$, $v_z = \partial\varphi/\partial_z$. If these are substituted in the differential equation of continuity, Laplace's equation emerges. On the other hand the streamlines are paths described by the so called stream function $\psi(x,y)$,

$$d\psi = \frac{\partial\varphi}{\partial y}\,dx + \frac{\partial\varphi}{\partial x}\,dy$$

We shall see somewhere else in this book that Laplace's equation is associated also with the velocity of sound in fluids. Under certain conditions, gravitational, electrostatic, magnetic, electric and velocity potentials satisfy Laplace's equation, which makes it an important tool of wide usage. It was on the basis of this formula that Laplace also corrected Newton's computation of the velocity of sound, i.e. showed† that the speed is $\sqrt{1 \cdot 4}$ times larger than was given by Newton. Indeed, if we consider gas to be made up of autonomous molecules, we find that the velocity of sound is of the same order of magnitude as the velocity of the molecules; and, according to the kinetic theory of gases,‡ the square of the velocity of the molecules is equal to $3p/\rho$, while the square of the speed of sound is $\gamma p/\rho$. Hence, the molecular velocity and sound velocity are in the ratio $\sqrt{3/\gamma} = 1{\cdot}46$, since $C_p/C_v = 1{\cdot}4$ for the air.

† *Aerodynamics*. Theodore von Karman, Cornell University Press, Ithaca, New York: 1954.

‡ See, for instance, *An Introduction to Molecular Kinetic Theory*. J. H. Hildebrand, Reinhold Publishing Corp., New York: 1963.

Augustin Louis Cauchy (1789-1857) and others

Cauchy occupies a special position in the history of Fluidmechanics, in several respects. In the first place, he was a man of great intellectual honesty, in the sense that he would not sacrifice his 'I', his convictions, for material advantages or formal honours. Like some of his contemporaries, he thought that *he who does not respect his own conscience will never be respected by others*. He could easily have become a prominent member of the Establishment of Paris (where he occupied three important professorships of mathematics) if only he had conformed with the state of affairs then in existence. But, no, when Cauchy was required, as anyone else, to take an oath of loyalty, he replied that it was an honour to remember one's duty but that *he could not carry two faces under one hood*.

Louis Philippe and his political bureaucrats did not like this at all, and so he had to leave his country to become an emigré first at Torino, Italy, then in Prague, where he taught mathematics. In 1851 he returned to Paris and, by special decree of the Emperor, was allowed to occupy a chair of mathematical astronomy without taking the oath of allegiance, thereby creating a shining example of academic integrity and behaviour. Before and after this, Cauchy published something like 180 papers and wrote 110 smaller items. Some of these were of great importance to Continuum Mechanics generally, and to Fluidmechanics in particular. It would, perhaps, be right to say that Euler and Lagrange created the subject. At the time of his death, Euler was working on a fundamental course of Hydrodynamics: Cauchy completed the elaboration and formation of some of its planned chapters. Laplace gave us the mighty equation (page 104); Cauchy showed that a similar equation existed for the stream function, $\psi = \psi(x,y)$. Laplace finalized the mathematical theory of the velocity potential: Cauchy developed it a step further, into the Cauchy-Stokes Theorem. Lagrange integrated Euler's equations of motion: Cauchy repeated this historic effort and arrived, in his own way, at the same integral which, therefore, is often called the Lagrange-Cauchy Integral.

In one of his most interesting papers,† Cauchy proved mathematically

† *Sur les dilatations, les condensations et les rotations produites par un changement de forme dans un système de points matériels*, vol. XII. Oeuvres completes de Cauchy.

that the motion of a fluid particle consists of a forward motion at the velocity $\bar{v}$ (v_x,v_y,v_z), a rotational motion with the angular velocity $\bar{\omega}(\omega_x,\omega_y,\omega_z)$, and a deformational motion characterized by the function $\Phi(x,y,z)$. The somewhat complex mathematical forms of this theorem can be found in many standard textbooks.† It was proved also by Hermann von Helmholtz, about whom more will have to be said later.

When one of the three velocity components is zero (for example $v_z = 0$), we have a two-dimensional flow. Consider, then, a two-dimensional potential flow for which $\rho =$ const (incompressible) and $v_x = \partial\varphi/\partial x$, $v_y = \partial\varphi/\partial y$. Euler's equation of continuity and Laplace's equation assume then the forms, respectively,

$$\frac{\partial v_x}{\partial x} + \frac{\partial v_y}{\partial y} = 0, \frac{\partial^2\varphi}{\partial x^2} + \frac{\partial^2\varphi}{\partial y^2} = 0$$

It can also be shown that the two functions – $\varphi(x,y)$ and $\psi(x,y)$ – are associated with each other through the so called Cauchy–Riemann conditions,

$$v_x = \frac{\partial\varphi}{\partial x} = \frac{\partial\psi}{dx}, v_y = \frac{\partial\varphi}{\partial y} = -\frac{\partial\psi}{\partial x}$$

or, what amounts to the same,

$$\boxed{\frac{\partial\varphi}{\partial x} = \frac{\partial\psi}{\partial y}, \frac{\partial\varphi}{\partial y} = -\frac{\partial\psi}{\partial x}}$$

It follows, then, that for an incompressible potential flow we have

$$\frac{\partial\psi^2}{\partial x^2} + \frac{\partial^2\psi}{\partial y^2} = 0,$$

that is, the stream function $\psi = \psi(x,y)$, too, satisfies Laplace's equation. Every solution of Laplace's equation may be taken as representing an irrotational (potential) flow. But to be of practical interest, the solution is additionally required to satisfy certain boundary conditions.

Georg Friedrich Bernhard Riemann (1826–66) was a German mathematician of outstanding abilities. He contributed to Mathematics in several lines: non-Euclidian geometry, the theory of functions, theory of potential, mathematical physics, mapping theorem, and so on. He was not a man of Fluidmechanics and his straightforward contributions to the subject were equal to zero. But some of his mathematical methods

† See, for instance, *Theoretical Hydrodynamics*. Kochin, Kibel and Roze, Interscience Publishers, John Wiley, New York: 1964.

were, and remain, very valuable, indeed. The Cauchy-Riemann equations, when transformed slightly, show, for example, that the lines along which the velocity potential function maintains a constant value, $\varphi = \varphi(x,y) =$ const (equipotential lines), and the lines along which the function $\psi = \psi(x,y) =$ const (streamlines), are orthogonal to each other. This means that any given two-dimensional flow of fluid can be represented by a certain distribution of the $\varphi =$ const and $\psi =$ const lines, as is shown in Figure 56 (*a, b*).

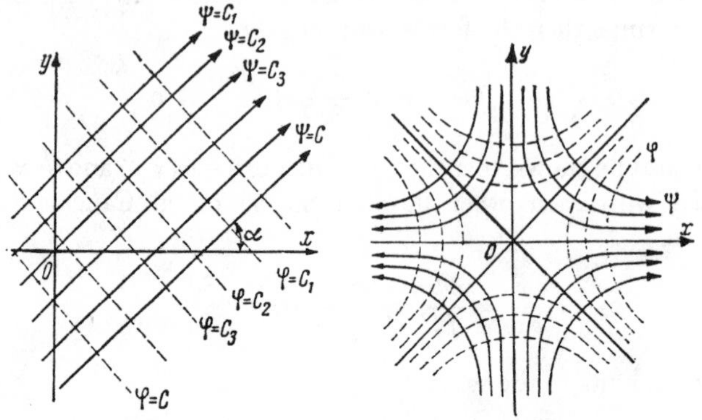

(a) (b)

Fig. 55. Streamlines and stream functions

Because the families of lines are orthogonal, geometrically it does not matter at all which of them are called streamlines and which equipotential lines. For this reason, φ and ψ are called conjugate functions. Thus, if we find a solution for some definite two-dimensional fluid-mechanic problem, i.e. if we manage to find a proper function φ (and, consequently, also ψ), we will find simultaneously a solution for another problem in which φ and ψ replace each other. And in this way we enter a very powerful branch of mathematics, the theory of analytic functions of the complex variable, which offers a whole range of interesting and important theoretical solutions of fluidmechanic problems, closely or by implications associated with names like Euler, Weierstrass,† Riemann,‡ Laplace, Cauchy, Zhukovsky, and others. Conformal trans-

† *Karl Weierstrass* (1815–1897). Werke, Bd. 3.

‡ *Riemann, Grundlagen fur eine allgemeine Theorie der Funktionen einer veranderlichen, Komplexen Grosse.* Werke, 2, Aufl., Leipzig: 1892.

formations, wing theory, sources, sinks, doublets and other techniques widely used in modern Fluidmechanics are based on this theory.

Hermann von Helmholtz (1821-94) and others

Undoubtedly, Hermann Helmholtz was one of the greatest scientists of the post-Leonardo da Vinci period. He was not just an anatomist, not just a physiologist, physicist and mathematician: no, he was also a philosopher, a great thinker interested in the mechanisms of interactions between the various natural phenomena and processes. Like Poiseuille, Jacobson, Boussinesq, Stefan, Ducleaux, Meyer and others, he studied fluid flows in relation to blood circulation, but his discoveries and inferences, at least some of them, proved to be of fundamental importance both to medicine and Fluidmechanics.

It was Helmholtz who pointed out that Girard's complaint that 'fluid particles refuse to move along straight lines' had its roots in the roughness of the surfaces of the pipes caused by the corrosive action of water upon copper.† Very much later on, J. N. Nikuradse, the well-known Caucasian aerodynamicist in Göttingen, showed experimentally that Helmholtz's theoretical conclusion was absolutely right, and that, indeed, roughness plays a very important role in the conditioning of pipe flows.‡

In common usage, by vortex we usually mean a whirlpool, or a circular cavity formed by a liquid in rotation. In the wake of this meaning, by vortex in fluidmechanics is meant a region of fluid bounded by the so-called Vortex Lines, whose tangents at all points are parallel to the local directions of the vorticity. By Vortex Motion of a fluid is meant its (fluid) motion with non-zero vorticity. The vector measure of local rotation in a fluid flow is called The Curl, or The Rotation, of the velocity vector $\bar{v}$, written Curl $\bar{v}$, or Rot $\bar{v}$. So, when Curl $\bar{v}$ is not equal to zero, there exists in the flow a string of rotating elements, a vortex line. And Helmholtz showed that these lines, which are the axes of rotation, have to be either closed curves, or they begin and end on the boundaries of the fluid or on the points in regions of infinite vorticity

† *Ueber Reibung tropfbarer Flussigkeiten, Sitzungsberichte K. K. Academie der Wissenschaften zu Wien.* H. Helmholtz und G. Piotrowski, B. XL, s. 656, v. 12, April 1860.

‡ *Gesetzmassigkeiten der turbulenten Strömung in glatten Rohren.* Nikuradse, J., Forsch. Gebiete Ingenieurw., 1932, Forschungsheft 356.

A vortex induces an external fluid motion, an external flow, so to speak. Figures 56 and 57, and Plate 13, show the pattern of such a flow. In the case of an ideal (inviscid) fluid, the external flow is in essence a field of velocities, an irrotational circulation, i.e. fluid particles move around the vortex, but they do not rotate about their own axes, therefore they have only forward velocities. Since these are caused by the vortex, they are called induced velocities.

Let us imagine now a purely circulatory flow, in which the particles move along circular streamlines without rotating around their own axes (Figure 56). Let the core of the vortex be like a solid body (this is an

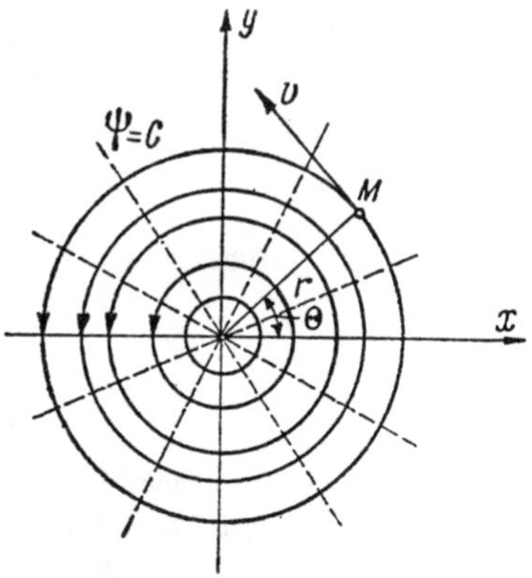

Fig. 56. Purely circulatory flow

accepted assumption in Fluidmechanics) so that its main kinematic characteristics are regular, free of the distortions implied by the Cauchy-Helmholtz theorem. The velocity field induced by such a vortex around itself is shown in Figure 57: each circular stream-line has a velocity u of its own. The 'work done' by u along the circular stream-line of length $2\pi r$ is called the circulation of u and is denoted by Γ, so that $\Gamma = 2\pi r u$. More generally, the circulation of any velocity v around any closed line $\Gamma = \int \bar{v}.d\bar{l} = \int v \cos\alpha dl$, where α is the angle made by v with the local direction of the curve, and dl is an element of the curve.

If the flow is potential, i.e. vortex-free ($\bar{\Omega} = 0$), and if the velocity potential $\varphi = \varphi(x,y)$ is a single-valued function of the co-ordinates (x,y), then $\Gamma = 0$ over any curve within the flow. Helmholtz proved in 1858 that the strength, or the moment, of the vortex is constant throughout its entire length, i.e. $\omega A =$ const everywhere (ω is the angular velocity, and A the cross-sectional area of the vortex). This is known as his second law, the first being the already discussed Cauchy-Helmholtz law. It is a purely kinematic law. But it can be shown that the invariability of

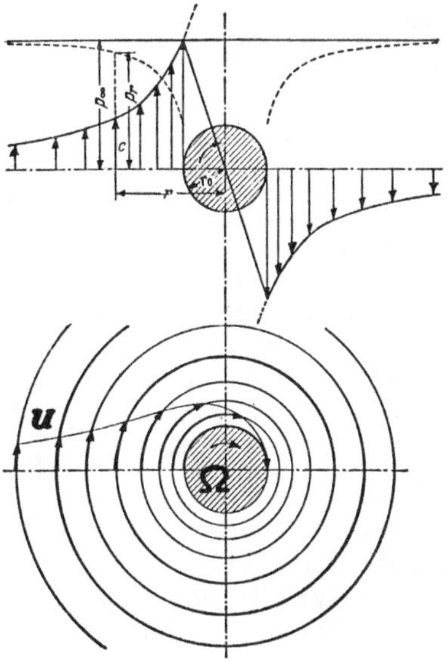

Fig. 57. The velocity field induced by a vortex

the moment of a vortex in time emerges also from the dynamical equations. The other laws, or theorems, by Helmholtz, are as follows: the moment of a vortex filament is invariable in time; every vortex line consists always of the same particles of fluid; particles that do not rotate at a given instant have never been rotating and never will rotate, so long, of course, as the assumed conditions hold.

These are, then, the historic theorems of Helmholtz. They express the indestructibility and the uncreatability of vortices in a non-viscous fluid which is acted upon solely by conservative forces, and of which the mass density is a function of the pressure alone.

I should, perhaps, add here that all these theorems have something to do with a more general theorem by Lagrange, who suggested, and proved, that if the fluid is ideal, the force acting upon a unit mass has a force potential, and if the mass density of the fluid is a function of pressure alone, then any part of the flow, which was vortex-free at the beginning, remains vortex-free. In other words, a flow of an ideal fluid, which was initially potential, will remain potential.

Lord Kelvin, or Sir William Thomson (1824–1907), a British mathematician and physicist, showed† that the circulation around any closed fluid filament, i.e. around any circuit composed always of the same fluid particles, is invariable in time. This means that if for a given instant of

† *On Vortex Motion*, Trans. Roy. Soc. Edinb. Sir William Thomson, Vol. 25, 1868.

time $\Gamma = 0$ for every circuit, then Γ will remain $= 0$ for all times. Expressed in yet another way, a motion of the fluid if once irrotational (potential) remains so for ever, or for as long as the assumed conditions continue to hold.

Sir George Gabriel Stokes (1819–1903) was an eminent British mathematician and physicist of Irish origin. During the years of his professorship at Cambridge, he made several important contributions to Fluidmechanics. For example, he derived the general equations of motion of real fluids in a rigorous manner, and so became the co-author of the Navier-Stokes equations (see page 90).

The so-called Stokes Flow is a flow of a viscous fluid at a very small Reynolds number when inertial, or acceleration, forces are negligible and the Navier-Stokes equations reduce to the simple form of $\mu\nabla^2\bar{v} =$ grad p. When the Reynolds number is less than 0·1, the fluidmechanic resistance of a sphere is given by the Stokes formula $R = 6\pi\mu rv$, r being the radius of the sphere and v its velocity.

We also have the celebrated Theorem of Stokes, which says that the surface integral of the Curl of a vector function, in our case Curl $\bar{v}$, is equal to the line integral of that vector function taken around the closed curve bounding that surface. To understand the fluidmechanic meaning of this theorem, let us recall from the preceding section that the 'work done' by the velocity vector $\bar{v}$ is called the circulation around the curve l, $\Gamma = \int \bar{v}.dl$. We have already seen that this expression is closely related to the vortex motion, and also to the most characteristic properties of irrotational (potential) motion.

It is important now to bear in mind that on inverting the sense of integration we change only the sign of the above integral, that is $\int_{ab} = -\int_{ba}$. To see this, let us consider carefully Figure 58. If acb is any conti-

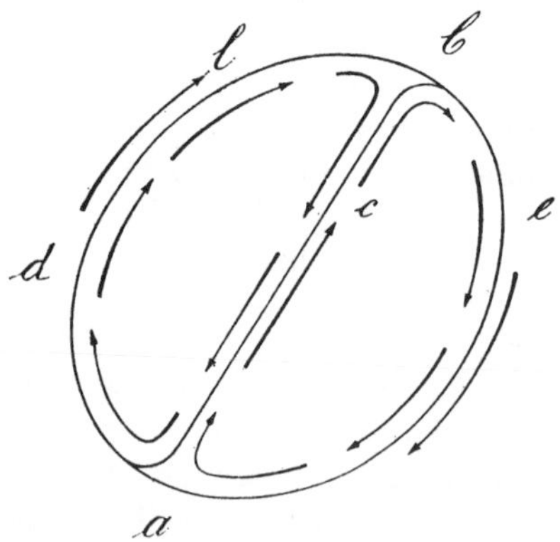

Fig. 58. A scheme illustrating Stokes' Theorem

nuous line joining a pair of points (a, b) of the circuit $l = adbea$, then $\Gamma_{adbca} = \Gamma_{adb} + \Gamma_{bca}$, $\Gamma_{acbea} = \Gamma_{bea} + \Gamma_{acb}$. But since $\Gamma_{bca} = -\Gamma_{acb}$, it follows that $\Gamma_{adb} + \Gamma_{bea} = \Gamma(l) = \Gamma_{adbca} + \Gamma_{acbea}$.

Similarly, decomposing any continuous surface bounded entirely by the given circuit l, as shown in Figure 59, we find that the 'work done'

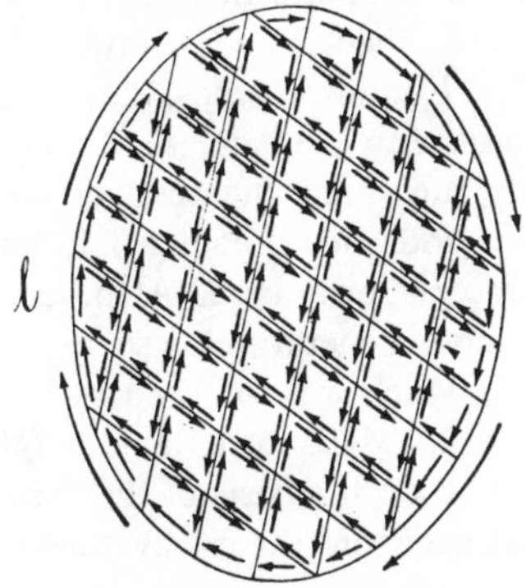

Fig. 59. Another scheme illustrating Stokes' Theorem

by the velocity vector $\bar{v}$ along the whole closed line l is equal to the algebraic sum of the 'work done' along the perimeters of each of the partial surfaces, and hence the Theorem of Stokes: the line integral of the vector $\bar{v}$ taken around the circuit l is equal to the surface integral of its Curl taken over any surface bounded by l. Or, in other words: the sum total of all the circulations around the small contours of the small surfaces (into which the surface has been divided) is equal to the circulation over the all-embracing contour l.

Needless to say, this was, and remains, a very important theorem. Its emergence was like the opening of the gates into a new castle of knowledge, like the conquest of yet another snowy mountain on the road to modern fluidmechanics.

We could go on and on naming more and more people who made valuable contributions to the shaping of Fluidmechanics as an academic discipline. But the number of pages at our disposal limits the list to just a few more names.

Gustav Kirchhoff (1824–87) of Germany and Baron Rayleigh (1842–1919) of Britain developed a theory of flow past inclined plates – a flow with discontinuity. This was supposed to correct both Newton's impact theory and d'Alembert's paradox (that the resistance of a body moving uniformly in a non-viscous fluid is zero if the fluid closes behind the body). But experiments and subsequent theories showed the high inaccuracy of the Kirchhoff-Rayleigh theory.

We have thus shown that up to about the end of the seventeenth and the beginning of the eighteenth centuries Aristotelean philosophy of scientific knowledge dominated the main streams of thought; almost all the leading pioneers of learning continued believing, for instance, that the time of fall of bodies was inversely proportional to their weights, i.e. that a heavy body falls faster than a light one. Galileo Galilei was the first to challenge Aristotle's theories. He showed, for instance, for the first time in the history of Fluidmechanics, that the resistance of a body moving in a fluid increases with the increase of its (fluids) mass density. While Mariotte actually measured the resistance – again for the first time. Sir Isaac Newton, and then Leonhard Euler, were the first to establish the historic $R \propto v^2$ formula. D'Alembert, as I have already indicated, advanced a more general yet also more rational theory of fluidmechanic resistance, which was followed by series of experimental investigations. William John Macquorn Rankine (1820–72), a Scottish civil engineer and physicist, on the basis of the theory of analytic functions, enriched the fluidmechanic theory by the so-called 'Method of Sources'.

Osborne Reynolds (1842-1912)

One major problem remained, however, unsolved. Namely, when the velocity of a body moving in a fluid is small, the fluidmechanic resistance is proportional to the velocity; therefore, its coefficient is reversely proportional to the velocity. When v is increased, the drag coefficient becomes almost independent of v; when v is increased still further, the drag coefficient drops to a smaller value. But what are the reasons?

To formulate the answer, it is necessary, once again, to recall that in simple terms, a fluid is a substance in which the intermolecular distances are very small compared with the body of the fluid itself; and there exists friction between the particles. The magnitude (or strength) of the latter depends on the type of fluid, and is determined by Newton's formula $\tau = \mu(du/dy)$, where μ is the viscosity coefficient, dv is the differential velocity at which one layer of fluid slides over another, and dy is the differential distance between the layers.

Newton's formula embraces a wide range of fluids (water, air, etc.). But there are also in nature substances for which the shear stress τ depends not only on the velocity gradient $(\partial u/\partial y)$, but also on the strain; these substances are visco-elastic. Another type of fluid is one with plastic behaviour which is characterized by an apparent yield stress, i.e. it behaves as a solid until it yields, and then behaves like a viscous fluid. On the other extreme, we have the so-called dilatant fluids, or simply dilatants, which flow easily with a low viscosity at low strain rates, but become more like a solid as the strain rate increases (Figure 60).

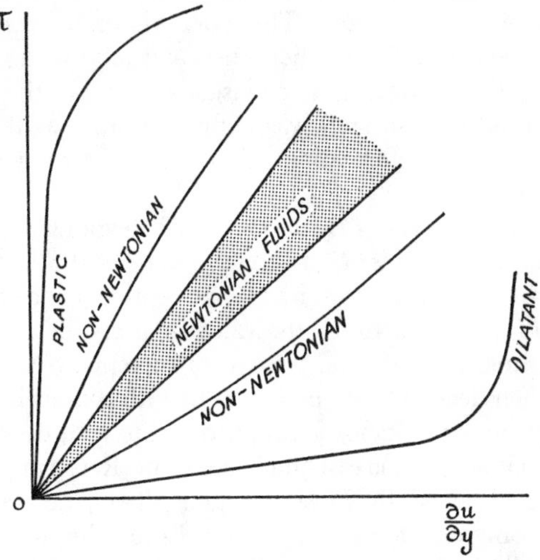

Fig. 60. Newtonian and non-Newtonian fluids

We can, of course, idealize water, air and gases as having no viscosity ($\mu = \tau = 0$), in which case they are referred to as ideal fluids. But in reality they are Newtonian viscous fluids. Which leads us to Osborne Reynolds: he carried out most thorough investigations of such fluids, and gave fluidmechanics a vast amount of new knowledge. His remarkable discovery in 1883 of the two modes of fluid flows, *laminar* and *turbulent*, became the starting point of important developments.

By the time Reynolds became a notable scientist, machines of all kinds had penetrated man's life. But how much do we know, asked Reynolds,† about those lubricants (and other substances) which are used to make machine surfaces smooth and slippery? He explained convincingly that there was still no consolidated theory for them, or for viscous fluids more generally. The reason, as he saw it, was that the problems of interaction between a solid surface and viscous fluid flow were far from being clear. Having studied Beauchamp Tower's work‡ on the subject, Reynolds

† *Papers on mechanical and physical subjects.* Osborne Reynolds, University Press, Cambridge, vol. 1 and 2, 1899, vol. 3, 1902.

‡ *Proceedings of the Inst. of Mechanical Engineers*, 1883–84.

came to the conclusion that at least some laws of lubrication could be derived from the already established hydrodynamic equations. He re-examined Newton's law and, to his surprise, discovered that he (even he!) knew very little about the velocity gradient (dv/dy). It was not in his nature to be satisfied with a little knowledge. He was soon able to say that this ratio can represent only one thing: deformation in the kinematical structure of the lubricant. Therefore viscosity was to Reynolds nothing other than shear force which exists and acts during the process of deformation of the lubricant, of the viscous flow, i.e. the coefficient of viscosity (or absolute viscosity) was defined by him as the ratio of the shear to the rate of deformation. Thus, if the rate was (u/a), then $f = \mu(u/a)$.

But is μ constant everywhere in the fluid? Reynolds asked. Will it be different for different fluids? Can it be the same at the surface of a solid body and at some distance from it? No, he replied. Since fluidmechanic resistance is proportional to v^2, the value of μ can be constant only in flows free of Stokes' eddies and transverse currents.

Reynolds then declared his disagreement with those who thought that when fluidmechanic resistance is proportional to v, μ depends on (u/a), and when (u/a) is small, μ is essentially constant. Referring to the experimental results obtained by Poiseuille and others, he asserted that μ depended not only on the ratio (u/a) but also on the dimensions of the pipe. The reason is, he continued, that the kinematic structure of the flow (in a pipe) changes from straight lines parallel to the axis into zigzag lines, from laminar into turbulent flow.

This discovery had many consequences. By his method of coloured bands,† Reynolds proved conclusively that in the first case there was a continuous steady flow of particles, in which, if the velocity were not time-dependent, the motion at fixed points remained constant, while, in the other, owing to small scale but intense eddy formation, the motion at fixed points varies according to no definite law.

Reynolds himself came to the conclusion that the conditions tending to stability and steadiness of motion were: an increase in viscosity μ; converging solid boundaries; free (exposed to air) surfaces; curvature of the path, with the greatest velocity at the outside of the curve; and a reduced density of the fluid. The reverse of these conditions, according to him, tends to produce turbulence.

The experiments by which Reynolds demonstrated the two modes of flow of water were carried out in glass tubes of various diameters up to 2 inches, and about 4 ft 6 inches long. These were fitted with bell-mouth entrances, and were immersed horizontally in a tank of clear water having

† Phil. Trans. Royal Society, 1883.

glass sides (Figure 61). The water in the tank was allowed to come absolutely to rest, and the valve *B* was then slightly opened, allowing the water to flow slowly through the tube. A little water, coloured with aniline dye, was then introduced into the mouthpiece through a fine tube supplied from the vessel *A*.

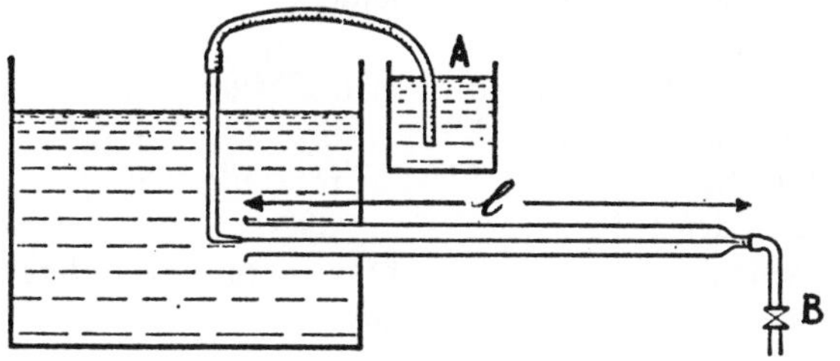

Fig. 61. Osborn Reynolds' experiment

At first this coloured water was drawn out into a single stream tube, extending through the whole length of the tube, the whole appearing to be motionless unless a light motion of oscillation was given to the water in the supply tank, when the stream line swayed gently from side to side, but without in the least losing its definition. As the valve *B* was further opened, the velocity through the tube increased, and the stream tube was drawn out more and more, still retaining its definition, until at a certain velocity eddies began to be formed intermittently near the outlet end of the tube (Figure 62).

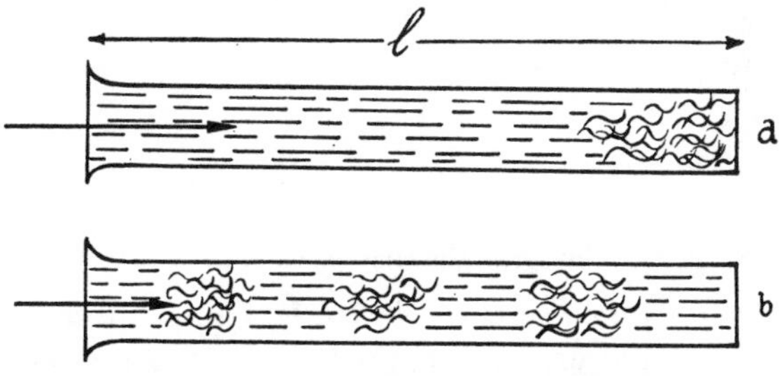

Fig. 62. Two modes of flow: laminar and turbulent

The formation of these eddies was accompanied by the almost instantaneous diffusion of the coloured band. As the velocity was still further increased, the point of initiation advanced towards the mouthpiece. The apparent lesser tendency towards eddy function near the inlet end of the tube was due to the stabilizing influence of the convergent mouthpiece.†
Finally, the whole flow became unsteady and turbulent.

† *Hydraulics and its Applications.* A. H. Gibson, Constable & Co. Ltd, London: 1908.

Any initial disturbance of the water tended to reduce the velocity at which the flow changed from steady to turbulent; this velocity was termed by Reynolds the Critical Velocity. Below and above this characteristic velocity there were Lower Critical Velocity and Higher Critical Velocity. The first of these represented the moment eddy formation was first noted, and the second the moment at which the eddies in flow originally turbulent died out.

The determination of the lower critical velocity proved to be beyond the capabilities of Reynolds' experimental techniques. But he took advantage of the fact that the law of resistance changed at the critical velocity and determined the values by measuring the loss of head accompanying different velocities of flow in pipes of different diameters. On plotting a curve showing velocities and losses of head (Figure 63),

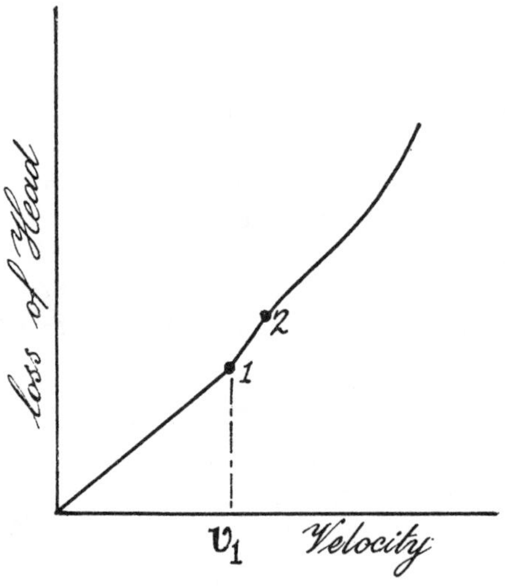

Fig. 63. Reynolds' diagram showing velocities and losses of head

it was found that up to a certain velocity, v_1, for any given tube, these points lie on a straight line passing through the origin of co-ordinates. From 1 to 2 there is a range of velocities over which the plotted points are very irregular, indicating general instability, while for greater velocities the points lie more or less on a smooth curve, indicating that the loss of head is possibly proportional to v^2.

Reynolds concluded that the critical velocity is inversely proportional to the diameter d of the pipe, and is given by

$$v_{cr} = \frac{1}{b}\frac{P}{d}$$

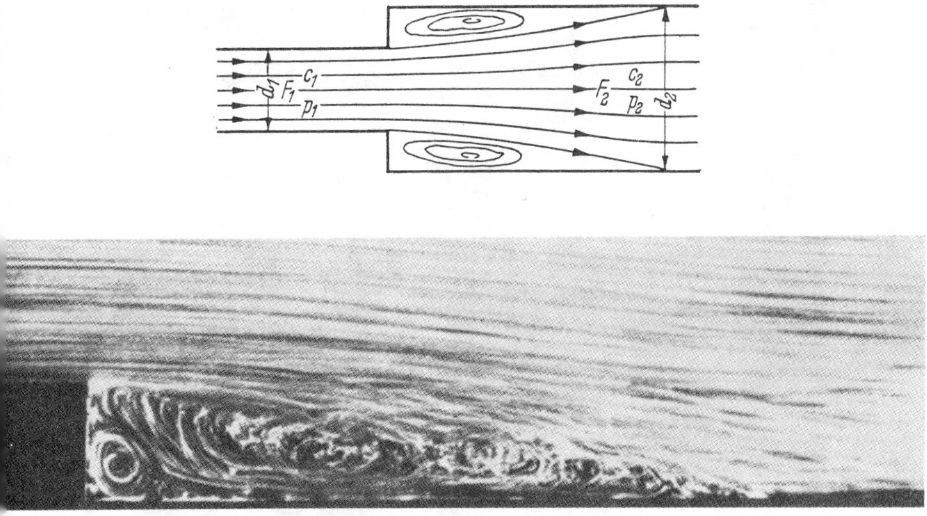

Plate 1. Borda's Theorem

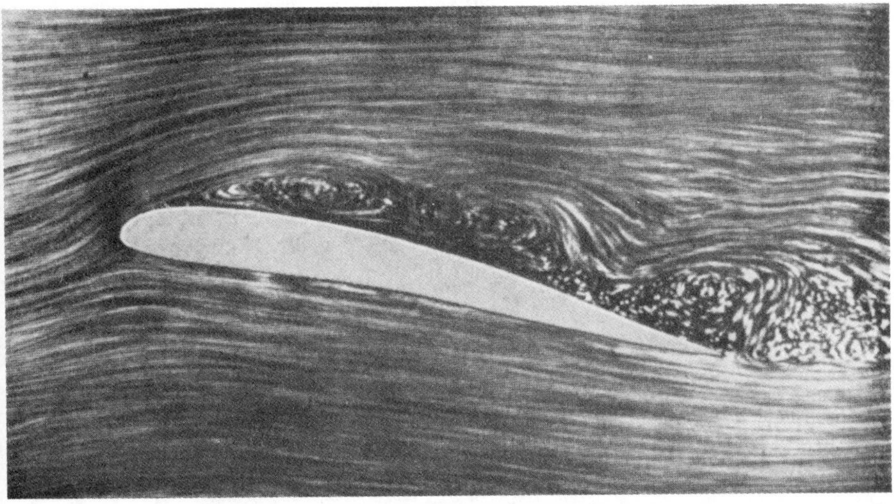

Plate 2. Flow separation on the upper surface of a wing

Plate 3. The tip vortices and vortex sheet behind an aeroplane in flight

Plate 4. Flow past a flat plate perpendicular to the flow

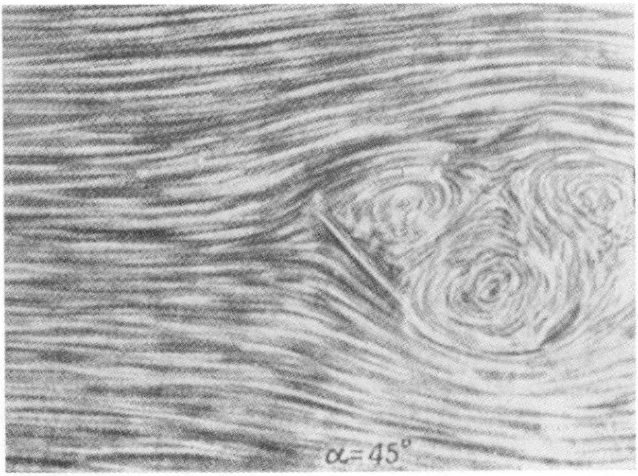

Plate 5. Flow past an inclined flat plate

Plate 6. Flettner's full-size Rotorship

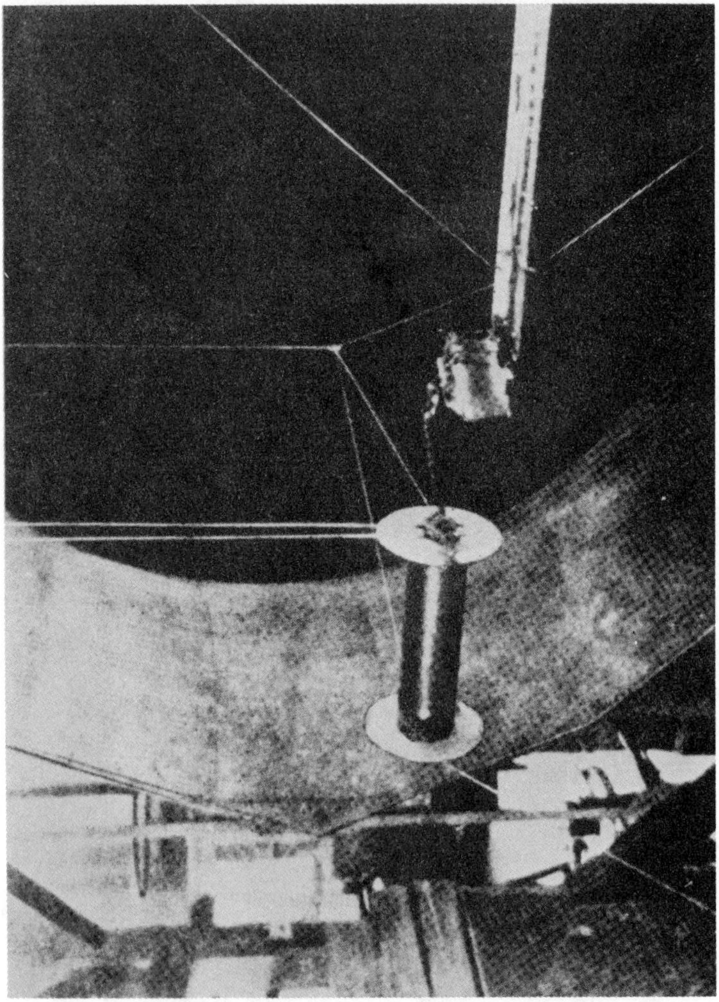

Plate 7. Flettner's rotor in a Göttingen wind-tunnel

Plate 8. Flettner's experimental Rotorship

Plate 9. Flettner's Rotormill

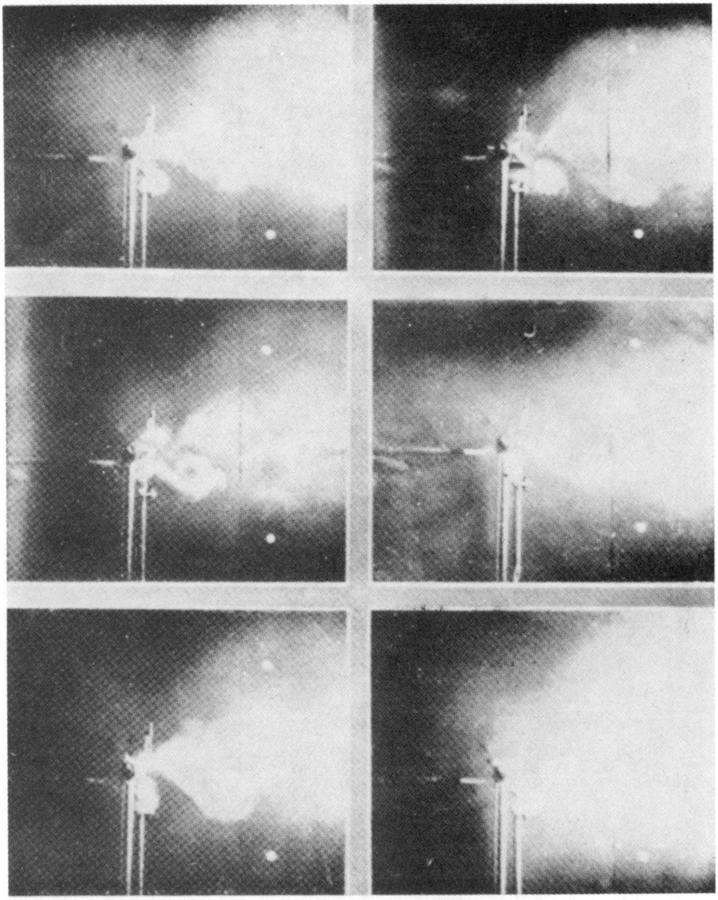

Plate 10. Riabouchinsky's flow patterns behind autorotating flat plates

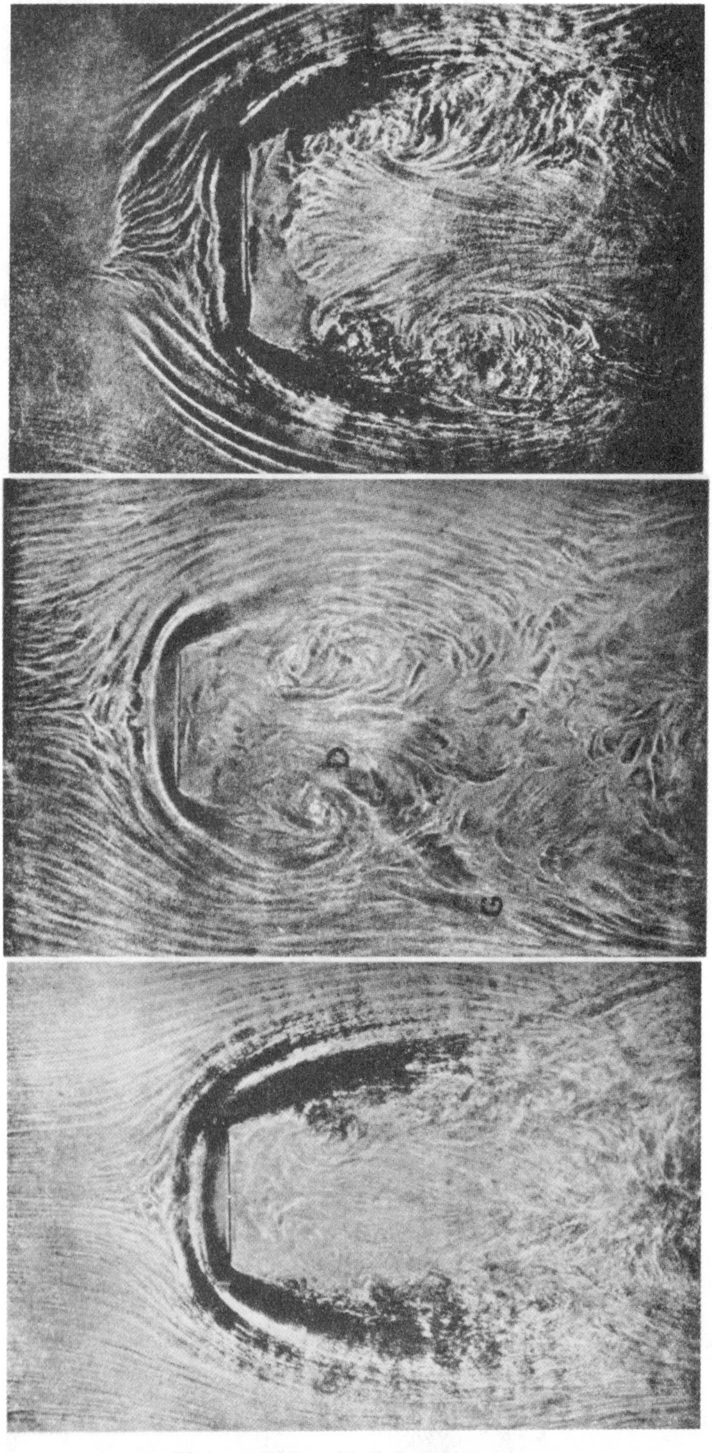

Plate 11. Riabouchinsky's flow patterns

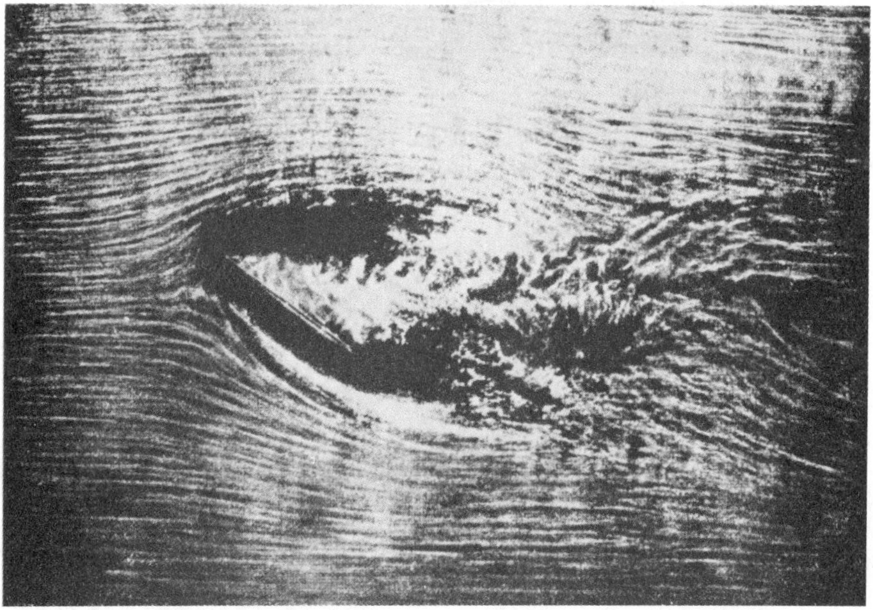

Plate 12. Riabouchinsky's flow patterns

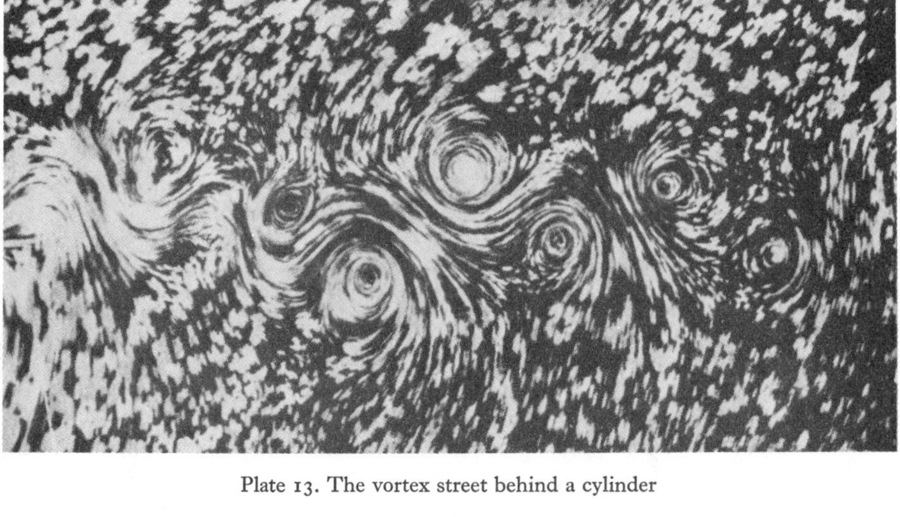

Plate 13. The vortex street behind a cylinder

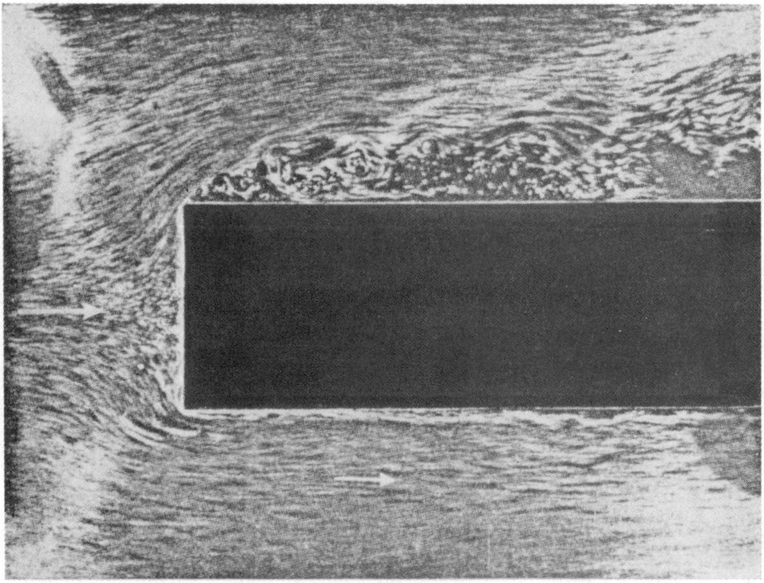

Plate 14. Flow separation

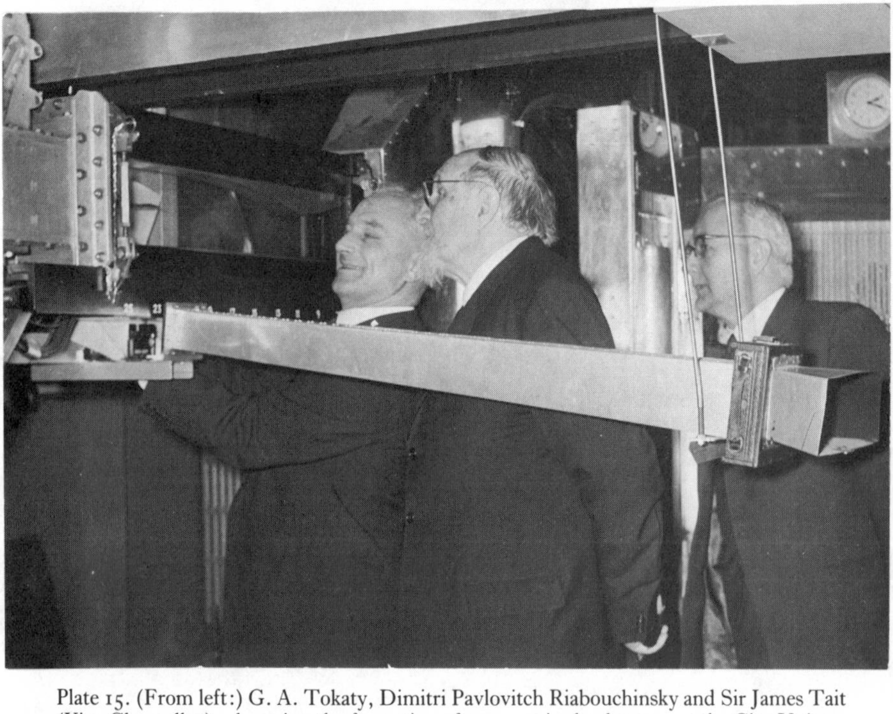

Plate 15. (From left:) G. A. Tokaty, Dimitri Pavlovitch Riabouchinsky and Sir James Tait (Vice Chancellor), observing the formation of supersonic shock waves at the City University, London, 1961

Plate 16. Riabouchinsky's moving ground actually constructed and in use

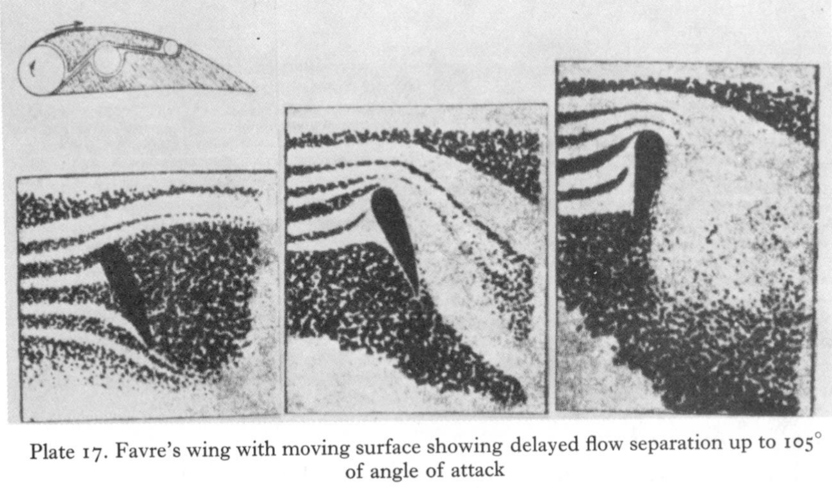

Plate 17. Favre's wing with moving surface showing delayed flow separation up to 105° of angle of attack

Plate 18. The vortex wake behind a propeller

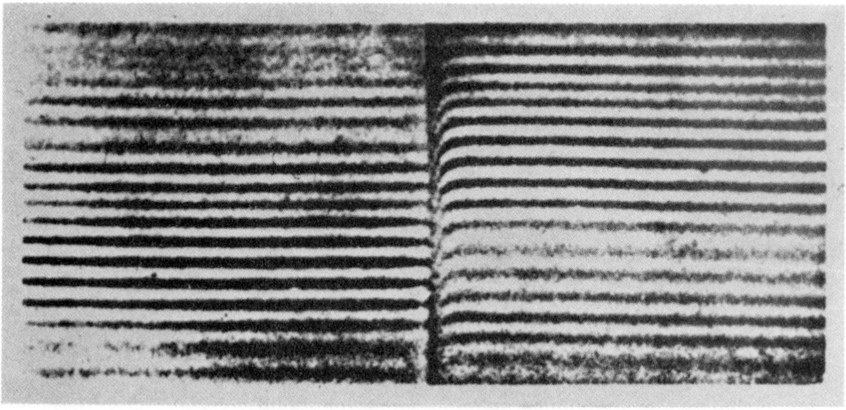

Plate 19. Photograph of the actual twist of the streamlines

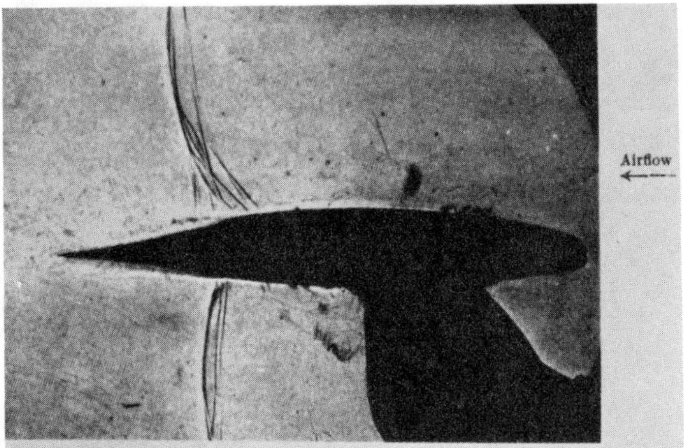

(a) Free-stream Mach number = 0·84.

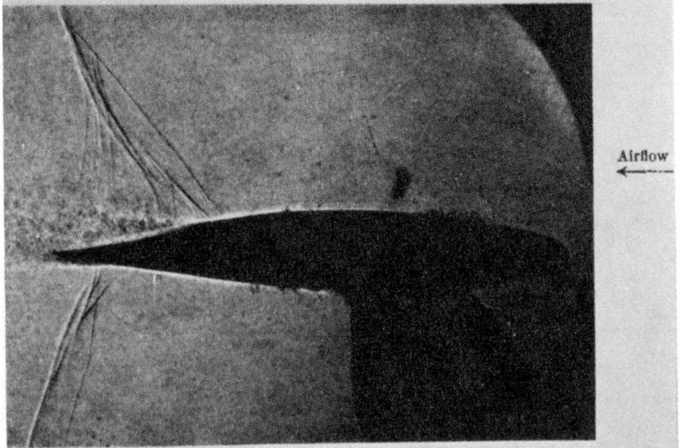

(b) Free-stream Mach number = 0·88.

Plate 20. (a) and (b) above and (c) on the following page. Patterns of shock waves at high subsonic Mach numbers (courtesy Pankhurst & Holder's 'Wind Tunnel Technique')

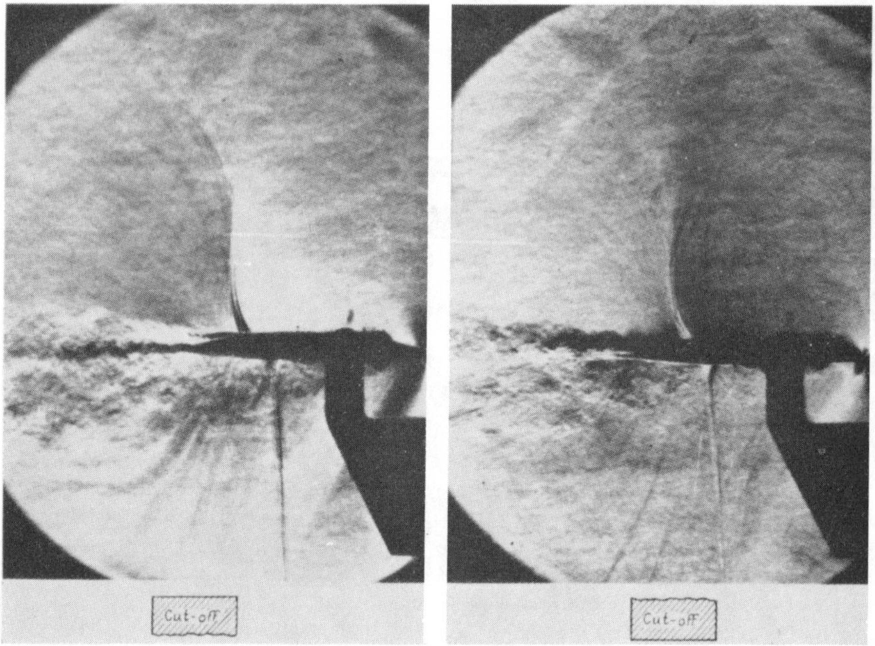

Plate 20. (c)

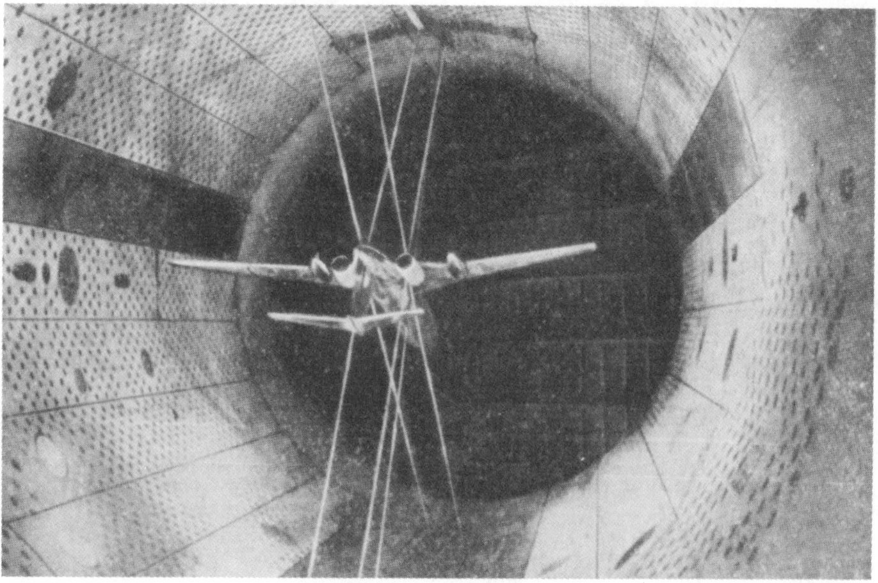

Plate 21. The working section of the large transsonic wind-tunnel of the CAGI of the USSR, with perforated walls

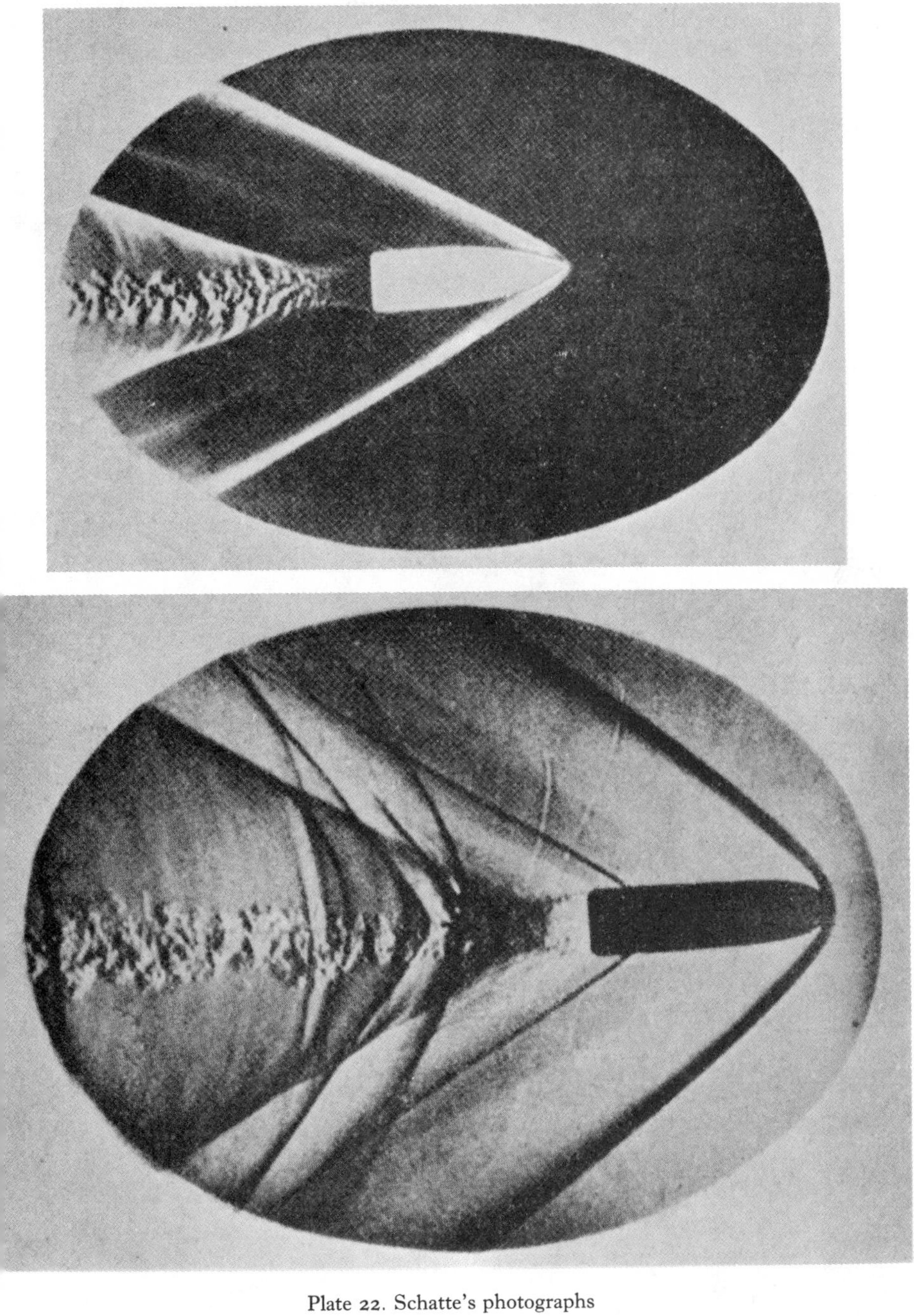

Plate 22. Schatte's photographs

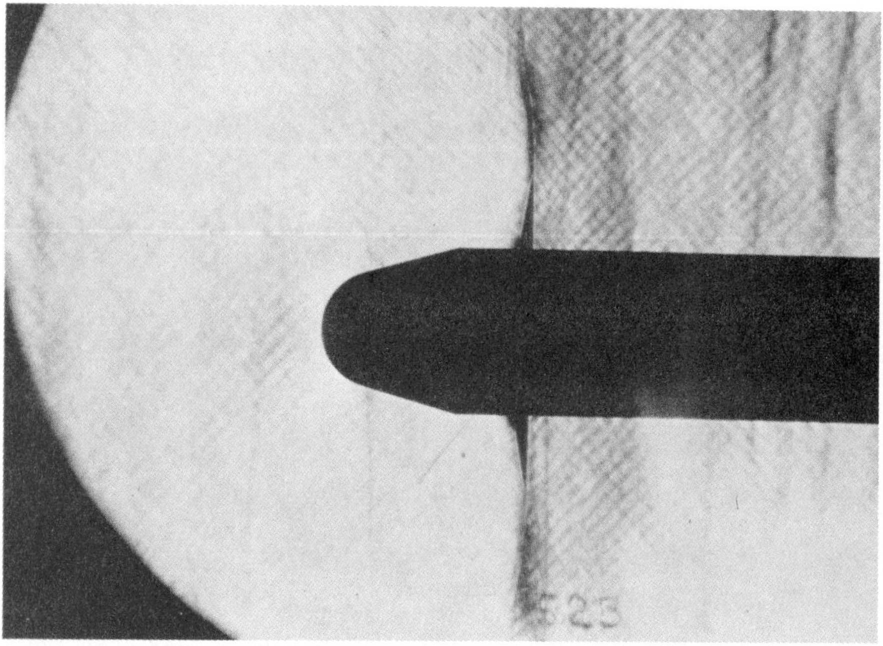

(a)

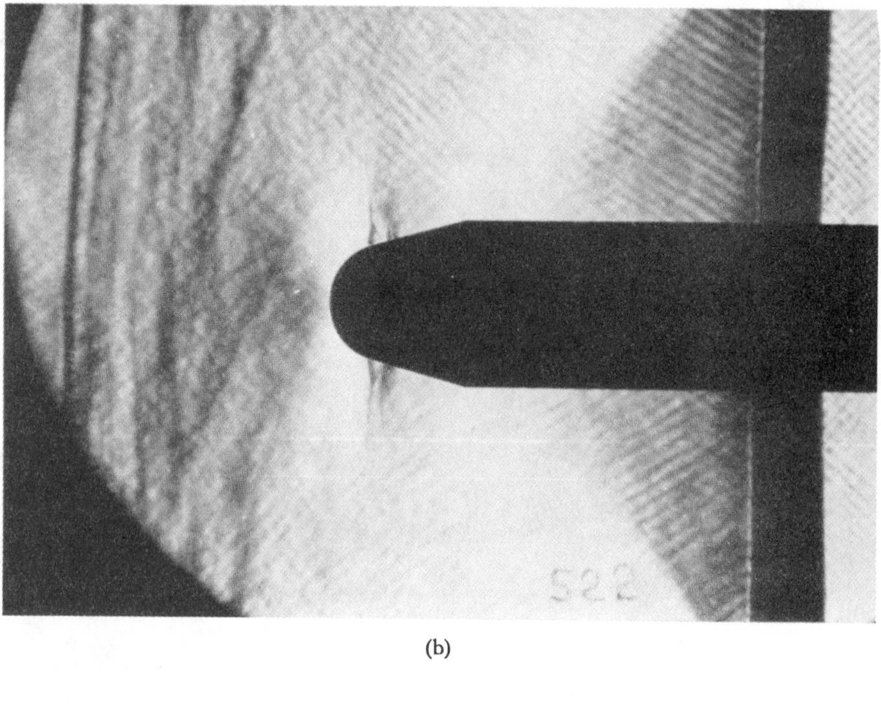

(b)

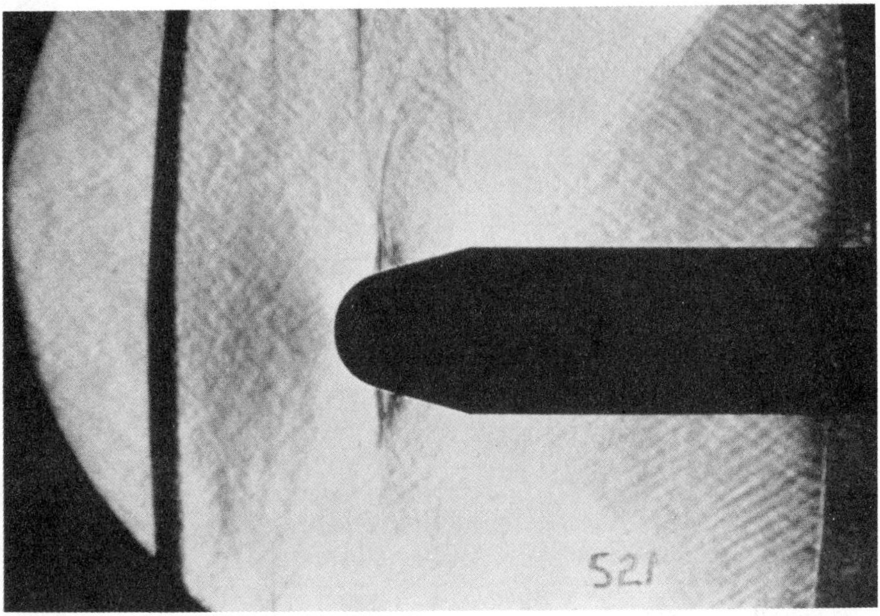

(c)

Plate 23. Shock waves obtained in the transsonic wind-tunnel of the Department of Aeronautics, The City University (M. M. Freestone):

(a) High subsonic flow from left to right ($M\infty = 0{\cdot}888$); shock waves emanate from the body (black) behind the shoulders

(b) Flow from left to right ($M\infty = 1{\cdot}029$); the transsonic shock waves can be seen emanating from the head configuration of the body (black)

(c) Flow from left to right; a typical case of an unsteady transsonic flow ($M\infty = 1{\cdot}051$), when the shock waves also are unsteady; ahead of the body (black) a normal detached shock wave in motion can be seen; and a complicated system of unsteady shock waves emanates from the head configuration of the body (black)

Plate 24. General view into the working section of the transsonic wind-tunnel of the Department of Aeronautics of The City University, London

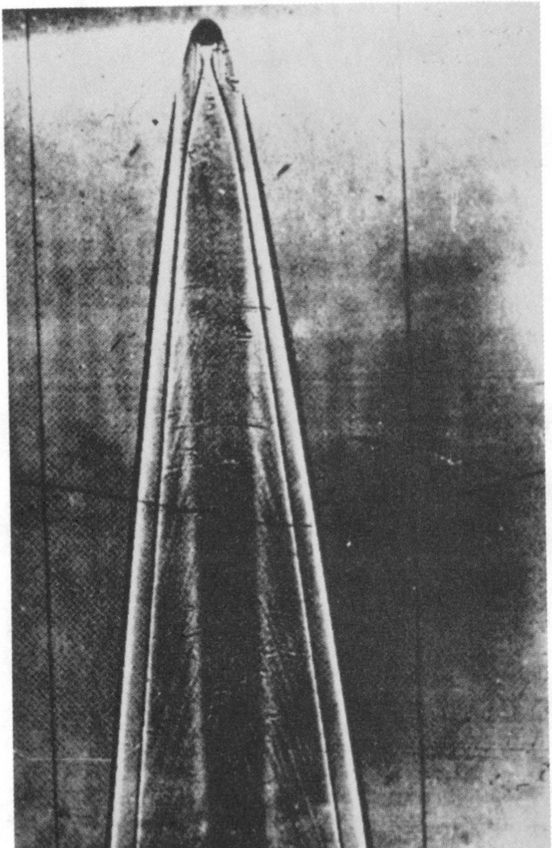

Plate 25. Hypersonic shock waves

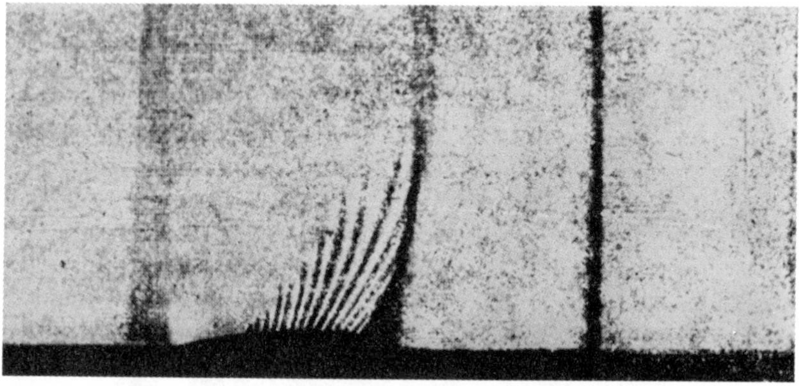

Plate 26. A fan of expansion shock waves on the wall of a wind-tunnel (flow from left to right)

Plate 27. G. A. Tokaty (left) and Theodore von Karman

where b is a numerical constant, such that if the length of the pipe is measured in metres, $b = 278$ for the lower critical velocity and 43·8 for the higher critical velocity. If t be the temperature in degrees Centigrade,

$$P = \frac{1}{1 + 0{\cdot}03368t + 0{\cdot}000221t^2}$$

A more general formula for the lower critical velocity in a straight pipe with parallel walls for any fluid and any system of units emerged in the form

$$v_{cr} = \frac{2000\mu}{d\rho}$$

Reynolds thus showed that μ measures the physical property of the fluid, which does not depend on its motion. He proved that in flows in straight pipes with parallel walls turbulence does not appear if the product of the mean velocity and of the diameter and of the mass density of the fluid divided by the absolute viscosity coefficient is less than a definite constant number, i.e. if $(\rho v d/\mu) < 1400$.

The ratio just given became known as the Reynolds Number,

$$\boxed{Re = \rho v d/\mu = \frac{vd}{\nu}}$$

Reynolds numbers are only comparable when they refer to geometrically similar flows; and then, provided all the other conditions can be described by the scale velocity and scale length, flows of the same Reynolds number are dynamically similar. Reynolds numbers are of great value in the various fields of Fluidmechanics, because tests of models are directly comparable to full scale results of geometrically similar shapes if the *Re* for the model is equal to the *Re* of the full-scale prototype.

It can be said that Osborne Reynolds revolutionized the whole concept of viscous fluid flows, injected into the theory of lubrication refreshing ideas and engineering techniques and, which is important, clarified the way towards the modern boundary layer theory. Indeed, it was he who came out with an unreserved declaration that the viscous fluid layer directly adjacent to the surface of a solid body is, because of viscous adhesion, pulled along at the speed of that body; and this was the starting point of the boundary layer theory created by Prandtl.

Mikhail Lomonossov (1711-65) and others

For whatever reason or reasons, many books on the history and philosophy of science and technology in the West either neglect completely or belittle the role played by the Russians (here and everywhere in this book 'the Russians' is used to mean all the peoples of the Soviet Union). As examples, but not necessarily the most striking ones, I could mention the following books: Réne Dugas' *A History of Mechanics* (Edition du Griffon, Switzerland, 1955); *The Road to Modern Science*, by H. A. Reason (G. Bell & Sons, London, 1947); *The Growth of Physical Science*, by Sir James Jeans (Cambridge University, 1947); *A Short Account of the History of Mathematics*, by W. W. Rouse Ball (Dover Publications, Inc., New York, 1960); *Inventors of our world*, by Joachim Leithäser (Weidenfeld and Nicolson, London, 1958). An inexperienced reader may well deduce from such books that the Russians invented nothing or little, contributed to knowledge nothing or little. But since Russia is far too big a cake to be belittled in this or any other way, we have to agree that omissions, probably with no ill intentions, become an East-West political issue.

A scientist should be a man or a woman determined to judge for himself or herself, on the basis of facts alone; he or she should not be biased by appearances, have no favourite hypothesis, be of no school, in doctrine have no master. 'Truth should be his primary object' (Michael Faraday). And when we survey the Russians from this summit, we see them in a different light, on a vastly greater scale.

To begin with, we must never forget that the giants of Fluidmechanics, Daniel Bernoulli and Leonhard Euler, although Swiss by birth and education, lived and worked in Russia as eminent members of the St. Peterburg Academy of Sciences. It is astonishing how few people in the West realize that very many of their pioneering theories were developed in that Academy.

We then turn to Mikhail Vassilyevich Lomonossov (1711–65), 100 per cent Russian and yet one of the greatest thinkers of his epoch. A poet and a historian, a physicist and a linguist, a chemist and a geographer, an inventor and an applied scientist, he can truly be called the father of scientific enlightenment in Russia. There has never been a shortage of evidence† of his outstanding contributions to knowledge. In 1748 he built a chemical laboratory which, among other problems, studied the physical nature of matter. One of the fundamental conclusions reached was that the division of matter into smaller and smaller particles was possible only up to a certain limit called 'atom'. Nothing new, of course: this proposition was known to man since the remotest antiquity. But Lomonossov carried out a rather deep study of the physical and chemical properties of this 'smallest particle'. The essence of his research emanated from the hypothesis that all substances are composed of elementary particles. According to this view, each physical body, including fluids, consists of extremely small bodies of matter which cannot be further divided physically and which are capable of mutual cohesion. In this terminology, they are 'insensible particles': by which is meant that they cannot be observed directly by a microscope or any other instrument.

Lomonossov stressed, however, that an 'insensible particle' remained a material body with such properties as mass, weight, volume, colour, taste, etc. I shall not discuss the general scientific-philosophical value of this concept, since it has already been discussed in chapters dealing with Aristotle, Archimedes, Euler, Leonardo da Vinci and others, but I would like to point out that the whole continuity theory of Fluidmechanics rests on it. We shall see, later on, also that the universal matter-energy continuity, too, would be impossible without such a foundation.

Lomonossov carried out a fundamental study into what became in due course a chapter in the Kinetic Theory of Gases.‡ We know today from textbooks of physics and thermodynamics that compression and expansion of fluids (gases) produce work: but that was precisely what Lomonossov wrote. We know that compressibility is one of the major peculiarities of gases: but that is precisely what Lomonossov proved both theoretically and experimentally. I do not suggest, of course, that no-one else worked in the field.

Why do gases get hot during compression and cold during expansion? he asked, and gave the answer: because in the first case the distances between the insensible particles become smaller, therefore the number of collisions increases, which gives rise to the temperatures of the gases; and in the second case the reverse is the physical truth. As to the motions of the particles (he said), they are of three types: translational ('progressive'), rotational, and vibrational. The warmer the gas, the greater

† We refer the English reader to the book *Russia's Lomonossov*, by Boris N. Menshutkin, translated from the Russian by Jeannette Eyre, Thal and Edward J. Webster, published in English by the Princeton University Press, USA: 1952.

‡ Coll. Works of M. V. Lomonossov, Moscow, 1934.

must be the rotational motion, the more strongly must the particles repel each other, therefore the greater is the compressibility of the air, the higher its pressure. Vice versa (he concluded), if the air were completely deprived of its thermal energy ('fire'), it would be entirely motionless – and here he repeated Voltaire's argument.

One contribution of Lomonossov is of particular interest and importance to the object of this book. He experimented with air kites and analysed the causes of winds. In 1751, he designed and constructed an anemometer for the determination of the strength of wind, and in 1754 designed and constructed a small meteorological observatory. The problem was how to use this equipment at various altitudes. And here Lomonossov showed brilliant inventiveness. This is what has been recorded in the minutes of the Peterburg Academy of Sciences:†

Esteemed Councillor Lomonossov demonstrated a machine he had invented and named 'aerodynamic machine' (Figure 64), its purpose

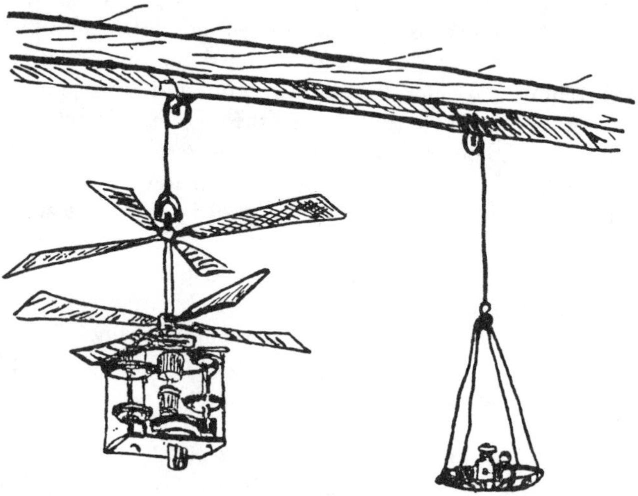

Fig. 64. Lomonossov's aerodynamic machine

being to elevate meteorological instruments into the upper layers of the atmosphere to investigate their condition and properties. The inventor reported that the wings of the machine are to be driven horizontally and in opposite directions by a clock-type spring. The machine was suspended at one end of a string passing through two pulleys, and a balancing weight was suspended to its other end. As soon as the spring was

† *Krylya Rodiny* (Wings of the Motherland). Lev Gumilevsky, p. 20, Moscow: 1954 (in Russian).

wound, the machine ascended to a height and thereby promised the desired action. This action, according to the inventor, will be even greater with a stronger spring and enlarged distance between the pairs of the rotating wings. . . .

Now, your excellencies, Lomonossov probably continued, what you have just seen with your own eyes, proves at least three things: (1) the air, which we do not see, is real enough to provide a support to the wings; (2) the design and construction of an aerodynamic machine is a practical proposition; and (3) the direct study of the upper layers of the atmosphere is no longer an academic dream.

It is up to the historians of aviation to decide whether Lomonossov was the creator of the first helicopter, and the inventor of counter-rotating propellers. I shall, in the meantime, present to you Mikhail Alekseyevich Rykatchev (1840–1919), a Russian professor of meteorology who, by the way, spent some time at the Greenwich Observatory. In 1870 he organized the first (in Russia) manned aerostat flights. In 1871, he carried out a series of experiments on the determination of the aerodynamic characteristics of lifting propellers. Moreover, he designed, constructed and tested a helicopter-like machine, whose four wings were rotated (in the horizontal plane) by a spring weighing 5·5 kilograms. Although there are no documental proofs of this, he is credited also with the autogyro idea. And it was he who warned his contemporaries that, however valuable in themselves, experiments alone would never make fluidmechanics what it could and should be.

The other great figure in the world history of science and technology is Dmitrii Ivanovich Mendeleyev (1834–1907), Russian chemist, the creator of the famous Mendeleyev Periodic Table of Elements.

He was an admirer of Joseph Louis Gay-Lussac (1778–1850) who, as is known, showed (1802) that different gases expand by equal amounts with rise in temperature. Two years later, he made a balloon ascension together with Jean Baptiste Biot, French physicist, and in due course made several trips up there on his own, one of which reached an altitude of more than three kilometres. No doubt this impressed Mendeleyev as much as the Gay-Lussac Law.

Lomonossov's helicopter, Rykatchev's experiments, Gay-Lussac's and others' efforts to penetrate the secrets of the atmosphere captured Mendeleyev's imagination to such an extent that he soon became an organic part of the Russian school of fluidmechanic thought. 'The prospect of knowing the laws governing the state of the atmosphere and the proud idea of aerodynamics captured me to such an extent', he wrote, 'that I put aside, for the time being, all the other activities and began studying aerostatics.' In 1887, he personally ascended in a balloon to an altitude of more than three kilometres. This was an exciting experience, he wrote, and I am now convinced that a method of exploration of the atmosphere has been found. But to make the method effective and safe, it is necessary to study the laws of air resistance.

His personal achievements in this field were numerous and significant. He studied the state of gases under a wide range of pressures, the theoretical and experimental properties of the upper layers of the atmosphere, hydrostatics, fluidmechanic resistance, etc. His book called *On the resistance of liquids and aeronautics* (1880) added his name to the list of the builders of Fluidmechanics.

The Russian School of scientific thought

Russian tsars and their governments, with the exception of Peter the Great, measured the greatness of their vast territorial empire by thousands of kilometers rather than by scientific and technological developments. Inadequately developed industry and commerce, very low level of literacy, lack of communications, etc. etc., kept thousands and thousands of talented men and women in isolation from the pulse of civilization. There were few schools and universities.

But still, Russia did produce men and women of genius. There were no laboratories to do experimental research, but nothing could prevent outstanding individuals from developing excellent theories. One of the great figures in the history of Fluidmechanics in Russia and the USSR, the late professor Vladimir Petrovich Vetchinkin (1888–1950), the author of the famous book on the mechanics of flight and of a large number of original articles, once said to us (his students) that: An Englishman could afford to build; the Russians could only afford to have theories. A Frenchman could afford to check his theories experimentally: the Russians could not go beyond their theories and self-made models. Therefore, Russia became primarily a country of theories rather than of theory and practice. . . .

Indeed, whatever aspect of life you take, the Russians have a theory for it and, as a general rule, a good one. I sometimes even wonder whether Russia was not next to France, at any rate next to Germany, in the development of the most fundamental theories of applied mathematics, mechanics and fluidmechanics. Yes, of course, they remained unknown to the outside world for far too long a time, probably because of the closed nature of Russian society, or because of the Russian language, or maybe, because of the widespread belief (in the West) that the Russians have always been only 1-metre tall.

When, in 1826, a young Russian mathematician, totally unknown, called Mikhail Vassilyevich Ostrogradsky (1801–62) appeared before the learned auditorium of the Paris Academy of Sciences to read a paper on heat conductivity, someone wrote (according to D. P. Riabouchinsky), or remarked: 'a Russian?, about heat conductivity?, what can he know of such a non-Russian subject?'. But when the lecture was over, someone else made a passionate speech in which he said that Fourier started the theory, Cauchy and Navier continued working on it, but Ostrogradsky completed it. As soon as Ostrogradsky's essay reached the thinking minds of the world, many of them agreed that, yes, there were – 'even in Russia' – men of Cauchy's, Fourier's and Navier's calibre.

Concerning the moments of forces (1838), *On the instantaneous displacements of systems in non-constant conditions* (1838), *On the principle of possible velocities and moments of inertia* (1842), *On the general theory of impact* (1857), *On the integrals of general equations of dynamics* (1850), *The differential equations related to the isoperimetric problem* (1850), *On the motion of a spherical shell in the air* (1841), *A course of lectures on celestial mechanics* (1836) – these were some of Ostrogradsky's publications.† We know from modern integro-differential calculus that there exists an Ostrogradsky's method of integration of rational function; and we know, of course, of the famous, if not revolutionary, Green-Ostrogradsky (or Ostrogradsky-Green) theorem, mathematically beautiful and of fundamental importance in Fluidmechanics, which says that the line-integral evaluated over a simple closed curve is equal to a double-integral defined over the interior of the plane region bounded by the curve. This theorem can be generalized into double-integral to triple-integral relationship, in which case it shows that if the *divergence and curl* of a vector field (for example, of a flow velocity field) are specified in some volume of space, the field is uniquely determined. That is to say, if, within a volume, the divergence and curl of a fluid flow are stipulated together with the value of the normal (to the surface) component, the field of the flow is uniquely determined.

The next outstanding Russian contributor to Fluidmechanics was Nikolai Dmitrievich Brashman (1796–1866), professor of the Moscow University. He studied the molecular structures of fluids, the stability of floating bodies, hydrostatics, and associated problems.

Then there was August Nikolayevich Davidov (1823–85), also professor of the Moscow University. In 1851, he completed a thesis dedicated to the theory of the capillary phenomenon and thus joined the

† *Evolution of Mechanics in Russia*. A. T. Grigoryan, Moscow: 1967.

ranks of those engaged in the study of fluids. He gave a detailed description of the conditions of equilibrium of bodies immersed in a fluid, illustrated the conditions geometrically, studied them mathematically, and produced practical recommendations. The Brashman and Davidov general line of investigation was continued by Russian professors: A. S. Ershov (1818–67) (*Water as a motive power*), I. I. Rakhmaninov (1826–1897) (*The theory of vertical water wheels*), I. V. Simov (1815–76) (*The analytical theory of the wave motion of the ether*), and many others.

Then there was Dmitrii Konstantinovich Bobylyev (1842–1917), probably the first teacher of Fluidmechanics in Russia, a man who is usually credited with the saying that ***no man should ever undertake experimental investigations until and unless they are demanded by a 'solid theory'.*** Fluidmechanic experts may be interested to know also that Bobylyev was the first in the world to study the theory of fluid flow past a wedge. He would not take anyone and anything for granted. Professor B. N. Your'yev once said of him: he used to say that a true scientist knows only if and when he himself, personally, proves or disproves.

By about the second half of the nineteenth century, Russia had become one of the leading nations in theoretical mechanics. The names of F. A. Sludsky, P. L. Tchebyshev, A. M. Lyapunov, I. A. Vyshegradsky, S. V. Kovalevskaya, N. E. Zhukovsky, V. A. Steklov, N. P. Petrov, D. S. Tehizhov, V. Ya. Tsinger, N. I. Lobatchevsky (1792–1856), I. S. Gromeka (1851–89), and many others, now represented the Russian school of thought in all its theoretical brilliance, with a heavy bias towards analytical methods and techniques: towards theory.‡

Alexander Mikhailovich Lyapunov (1857–1918) was a typical representative of that school. He worked as professor of Mechanics in the Universities of St. Peterburg, Khar'kov, and Odessa. In 1880, he undertook a mathematical study of the problem of equilibrium of heavy bodies in heavy fluids, and one year later published his first paper under the same name.† This was followed by his second publication under the title, *On the potential of hydrostatic pressures* (1881). In 1882, he completed his M.Sc. thesis called 'The stability of an ellipsoidal form of equilibrium of a rotating fluid'.

Although these were still only the first steps along a long road ahead, Lyapunov impressed his contemporaries at home and abroad by the originality and depth of his methods. A whole range of further papers published by him ('The general problem of stability of motion', 1892; 'The stability of spiral motions of a solid body in a liquid', 1888; 'The stability of motion of three bodies in a liquid', 1889), and special lectures

† *Selected Works*. A. M. Lyapunov, Moscow: 1948.

‡ Those interested in the subject may address themselves, for example, to *Lyudi of Russian Science*. Science Publishers, Moscow: 1965 (in Russian).

at Odessa University, made his name widely known all over the world, because the field he studied so successfully was of great importance not only in fluidmechanics, but also in general mechanics, ballistics, astronomy, theory of mechanisms, and mathematics.

I should now like to introduce to you a new figure of whom, I am sure, you have never heard; who is still unknown to the majority of the peoples even of the USSR itself, but who will soon receive from the grateful world of space science and technology full recognition, a profound appreciation: Ivan Vsevolodovich Meschersky (1859–1935). It is a matter of great personal honour and pride to me to say that I have met many giants of science in my life: Max Planck, Ludwig Prandtl, Theodore von Karman, Dimitri Riabouchinsky, Sergei Tchaplygin, Konstantin Tsiolkovsky, Albert Einstein, Aleksandr Kotel'nikov, Ivan Meschersky, and others.

Unless my statement is repudiated by someone, I am going to say that Meschersky, a graduate of the St. Peterburg University, then a postgraduate student under Professor Bobylyev, then Docent at that University, then, for over thirty years, Professor of Mechanics and Applied Mechanics at the St. Peterburg Polytechnic Institute, will be known to future generations as the father of Analytical Rocketdynamics. For it was he who developed the theoretical mechanics of bodies of variable masses:† that branch of mechanics without which there would be no theoretical rocketdynamics. To understand the general significance of his contributions, let us recall that the motor car, the aircraft, rockets, ships and other machines driven by combustion engines of all kinds are, really, bodies of varying masses, while Sir Isaac Newton's law of motion was derived for $m = \text{const}$. I might add that Meschersky's works are also of significant mathematical value.

Konstantin Tsiolkovsky (1857-1935)

And now to turn to Konstantin Eduardovich Tsiolkovsky, one of the most fascinating figures in the history and philosophy of science and technology, the acknowledged 'father of rocketry', the great Russian scientist well-known both in the USSR and abroad. In the course of his original research, he not only predicted the development of jet aircraft and long-range rockets but theoretically confirmed their feasibility. He was much preoccupied also with the problems of constructing metal

† *Works on the mechanics of bodies of variable masses.* I. V. Meschersky, Moscow: 1952 (in Russian).

dirigibles, and was a pioneer in the field of experimental aerodynamics. In 1897, in Kaluga, this self-educated provincial teacher built one of the world's first wind-tunnels (Figure 65), and for five years used it

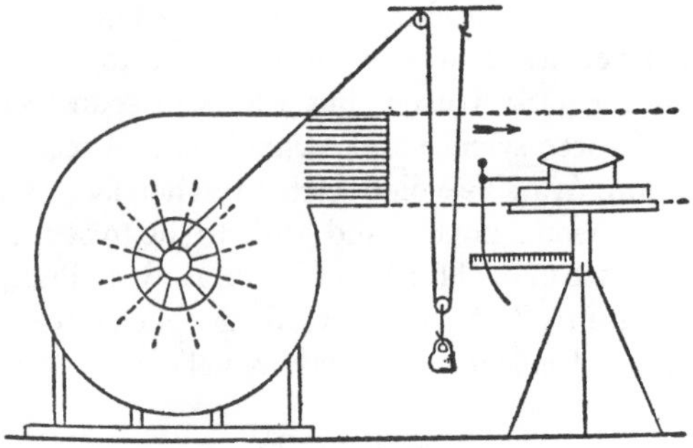

Аэродинамическая труба Циолковского.

Fig. 65. Tsiolkovsky's wind-tunnel

consistently to investigate models of airfoils, dirigibles, various geometrical shapes,† etc. His powers of analysis and theoretical reasoning enabled Tsiolkovsky to derive a number of important laws of aerodynamics.

In his *Aeroplan, ili ptitsepodobnaya letatel'naya mashina*, the airplane, or bird-like flying machine, published in 1894, Tsiolkovsky produced the first aerodynamic configuration of an aeroplane (Figure 66), a monoplane with unbraced streamlined wings, wheeled landing gear, and

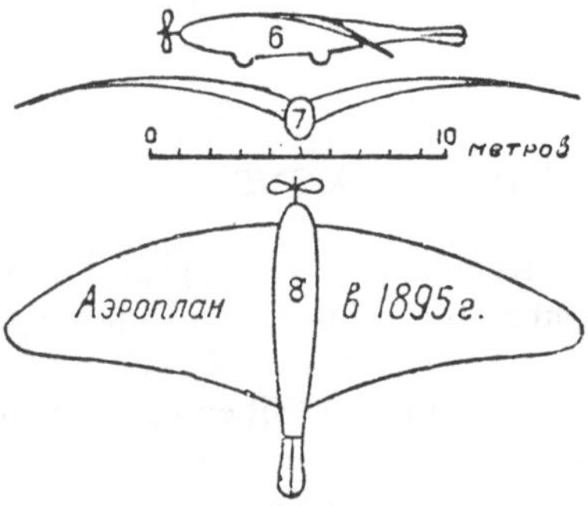

Fig. 66. Tsiolkovsky's aeroplane

† *Coll. Works of K. E. Tsiolkovsky*. The Academy of Sciences of the USSR, Moscow: 1951 (in Russian). Available also in English, NASA TTF-236.

coaxial rotation of the propellers. Generally, the problems of aerodynamics held considerable interest for him throughout his many-sided life. His first work connected with the problems of aerodynamics (*The Theoretical and Practical Aspects of an Aerostat with a shape elongated in the horizontal direction*) was published in 1885 and his last (*The pressure on a plane in a normal flow*) appeared in 1934.

For Tsiolkovsky aerodynamics was not merely of theoretical interest. He clearly understood that without reliable knowledge concerning fluid-mechanic resistance no progress would be possible in aeronautics. In his own work, he always subjected his theoretical results to experimental verification. Whenever he introduced an assumption, it was always with the reservation that this assumption and the conclusions derivable from it were only valid to the extent that they could be justified by experiments – and he himself was an indefatigable experimentator. He never left his experimental results in the form of a random accumulation of facts; he always sought and, with striking intuition, invariably found general theoretical relations between the quantities observed.

In 1890/91, Tsiolkovsky wrote a paper, *On the problem of winged flight*, in which he employed completely original methods of study of the air pressure on a flat plate. It is interesting to observe that, kinematically and dynamically, this work was similar to the already discussed Newtonian-Eulerian impact theory. Tsiolkovsky, however, was not in the habit of repeating anyone's ideas blindly. Moreover, the work in question is full of evidence that he arrived at his propositions independently, in a typically Tsiolkovsky manner. Like Newton and Euler, he concluded that the fluidmechanic resistance of an inclined flat plate is proportional to the square of the sine of the angle of attack and, all the other things being equal, to the square root of the aspect ratio of the plate. This remarkable law anticipated the development of the airfoil theory by about thirty years.

For the experimental verification of his theoretical concepts, Tsiolkovsky invented a simple device, a pair of wings mounted on a shaft which could be rotated for a certain time by means of a string unwinding under the action of a weight. By measuring the latter, and by the amount of upward movement of the wings in the direction of their axis of rotation, Tsiolkovsky determined approximately the lift and drag of the wings. 'With sparing use of analysis and simple apparatus, Tsiolkovsky arrives for the most part at the right results', wrote the great N. Ye. Zhukovsky: 'His original method of investigation, his reasoning and ingenious experiments are not without interest and, in any event, mark him out as a talented investigator'.

Tsiolkovsky studied the mechanics and aeromechanics of birds' flight.

But (he wrote) man will never be able to fly 'on his muscles alone', because 'the imitation of birds is technically very difficult owing to the complexity of the movements of the wings and tail and the equally complex co-ordination of their motions'. Therefore (he continued), I suggest a flying machine that has the shape of a rigid, soaring bird, with two propellers at its head rotating in opposite directions and developing a forward thrust. . . . In such a machine, the role of the muscles will be played by an internal combustion engine. . . . Instead of the tail of a bird, there will be a dual surface, one for vertical and the other for horizontal control. . . . He then analysed the take-off and landing characteristics of the machine and suggested, probably for the first time in the history of aviation, a method of performance analysis: the speeds required, engine power, etc.

In 1894, Tsiolkovsky designed yet another experimental aerodynamic facility, to study the comparative drag characteristics of bodies of various configurations. It was a pivoting arm with the test model attached to its end. He soon discovered, however, that this and similar devices could not give him satisfaction, and he came to the conclusion that, instead of moving test models in the stationary air, air should be made to move past stationary models. Hence, his idea of a wind-tunnel (see Figure 65). This is what he actually wrote: 'Recently, while conducting certain experiments, I got the idea of using a completely new method involving an artificial wind (created by a vaned blower). . . . So far I have experimented only with a model 42 cm long. The experiments confirmed our formulae, and the drag coefficients, which I obtained for this model, gradually decreased with increase in the velocity of the artificial air flow. The new method makes it possible to carry out investigations at any time and with considerable accuracy; it is also very convenient for demonstration purposes.'

In 1898, Tsiolkovsky published† an article called 'The Air Pressure on Surfaces in an Artificial Air-Flow' which can truly be called the actual beginning of experimental aerodynamics, in the sense of the definition given at the beginning of this book. At the end of the article, he wrote that the correct formulation of the laws of aerodynamic resistance was very important, but that it cannot be achieved without experiments. Also: 'Experiments make an enormous contribution to the theory of the aerostat and the aeroplane; in fact, is there any field of science and technology in which the laws of fluidmechanic resistance play no part? And this underlines the importance of experiments in Fluidmechanics.'

In 1899, a member of the Russian Academy of Sciences, M. A.

† *Collected Works*. Vol. 1, Moscow: 1951. Also his *Selected Works*. MIR Publishing House, Moscow: 1968 (available in English).

Rykatchev, a distinguished aerodynamicist himself, reported to the Academy that Tsiolkovsky's experiments were 'more than just experiments', because they rested upon 'solid theoretical foundations'. In spite of the primitiveness of his home-made wind-tunnel (continued Rykatchev), Tsiolkovsky advanced experimental aerodynamics to the level of science and made it a respectable branch of knowledge. It is, therefore, desirable to give him a chance to continue his experiments on a larger scale, with better measuring instruments. Although the Academy was rather proud of its theory-first tradition, and although there were in it influential figures who thought that knowledge accumulated experimentally was neighbouring with intellectual vulgarity, Rykatchev's recommendation was accepted, Tsiolkovsky was given a good chance to continue his pioneering investigations. By 1900, he had designed and constructed a new wind-tunnel and associated measuring devices. Once again, he proved by deeds that Russia had in him an outstanding builder of Experimental Aerodynamics.

His still more important contributions were, however, in the field of rocketry. I had the honour of meeting him (1932–5), knowing him, learning from him; I am proud of being one of his followers in rocketry; it seems to me that I have read every word and studied every formula and drawing published under his name. Therefore I should like to allow myself to describe Tsiolkovsky as one of the greatest empiricist-inventors with the lowest formal education in the history of aeronautics and astronautics.

Nikolai Egorovich Zhukovsky (1847-1921)

Professor Zhukovsky, the official 'father of Russian aviation', was a towering figure of a different academic and educational background. Tsiolkovsky was a mere provincial school teacher; Zhukovsky was an eminent professor of mechanics in an eminent higher educational establishment. The first lacked adequate thoretical knowledge, the second was one of the brilliant creators of that knowledge.

In 1868, Zhukovsky graduated from the faculty of Mathematics and Mechanics of the Moscow University. During the next two years he continued his study at the St. Peterburg Institute of Railway Engineers. In 1870–2 he taught physics at a gymnasium. In 1872 he started teaching mathematics at the Moscow Higher Technical College (MVTU), in 1874 he became Docent in analytical mechanics at Moscow University. From then on, he climbed the academic ladder steadily and successfully, year after year, becoming more and more known at home and abroad.

It would, indeed, be difficult to name a branch of mechanics beyond his interest. I shall, however, restrict my description to his role in Fluidmechanics. Zhukovsky's first contribution was his M.Sc. dissertation on the *Kinematics of a fluid body* (1876, Moscow University). In it, he gave a beautiful elaboration of the geometrical and physical characteristics of an ideal fluid flow. The formula 'motion of a fluid particle' became much more meaningful. This was followed by series of articles on the theoretical mechanics of solid bodies† of various types: impact of two bodies (1885), the theory of the gyroscope (1895), pendulum theory (1895), theory of rotation of a solid body (1892), stability of motion (1882), the motion of a body containing fluid (1885), etc., etc.

On 24th April, 1881, Zhukovsky delivered a public lecture on the aerostat (balloon) problem; on 12th February, 1882, he addressed the Society for the Advancement of Mathematics and Physics on 'A machine for the solution of equations', and on 30th April of the same year, on the 'Reaction of fluids'. On 4th February, 1883, he gave yet another public lecture on the determination of the orbits of plants and comets, and from 18th to 28th April, in Odessa, three further lectures: 'The oscillations of floating bodies', 'The impact of two spheres, one of which floats on the surface of water', 'A method of solution of the main equation of motion of planets'. On 27th December, 1884, Zhukovsky appeared before the Society of Mathematics and Mechanics with a new topic: the role of the surface of a body moving in a fluid in the formation of the fluidmechanic resistance. On 10th January, 1886, Zhukovsky addressed the Moscow University's Society of Mathematics on 'The Speed of sound' in fluids, especially in the air. In June of the same year he was elected professor of Mechanics at the University.

Yes, Zhukovsky emerged as the Lagrange of Russia. His scientific reputation soared very high. Subsequent history showed, however, that all these were no more than the first glories. The fluidmechanics of subsoil flows; the kinematics and dynamics of vortex rings; the basic theory of flight (1890); on the Helmholtz theory of streams; a modification of Kirchhoff's method for the determination of a two-dimensional fluid flow; on Otto Lilienthal's‡ investigations; on the soaring of birds;

† *Collected Works*. N. E. Zhukovsky, vols I–VII, Moscow: 1948–50 (in Russian).

‡ Otto Lilienthal (1848–1896), German aeronautical engineer, who studied the flight of birds and built gliders in which he demonstrated the advantages of curved surfaced wings over flat wings. A friend of Zhukovsky. Killed during a flight.

on the fluidmechanic resistance of boats; a new method of study of motion of bodies in water; and so on, and so forth: all these made N. E. Zhukovsky by about 1900 not only one of the most productive men in the history of Fluidmechanics, but also one of its most original and effective builders, at least in Russia.

In a book of this nature, it is necessary, however, to single out one or two major contributions capable of showing the role and place of their author in history. Torricelli is known mainly by his Torricelli's Law, Bernoulli by Bernoulli's Law, Euler by his Euler's Equations of Motion, Newton by his laws of motion, Lagrange by his Lagrange's Integral, d'Alembert by his d'Alembert's Paradox, Reynolds by his Reynolds Number, etc. What are, then, the one or two major contributions by Zhukovsky which put him in the forefront of the fathers of Fluid-mechanics?

There are several such contributions: Zhukovsky's conformal transformation, Zhukovsky's aerofoils, Zhukovsky's hypothesis, etc. The flow around any given aerofoil may be described by the appropriate complex potential $\xi(z)$; the simplest aerofoil is the Zhukovsky aerofoil, obtained from the flow about a circle by a single conformal transformation, the main instrument being the Zhukovsky Transformation Formula $\xi = z + a^2/z$. Namely, substituting this formula into the expression for the flow about a circle of radius slightly larger than a, and so placed that the circumference passes through the point $\xi = a$ (Figure 67), an aerofoil configuration

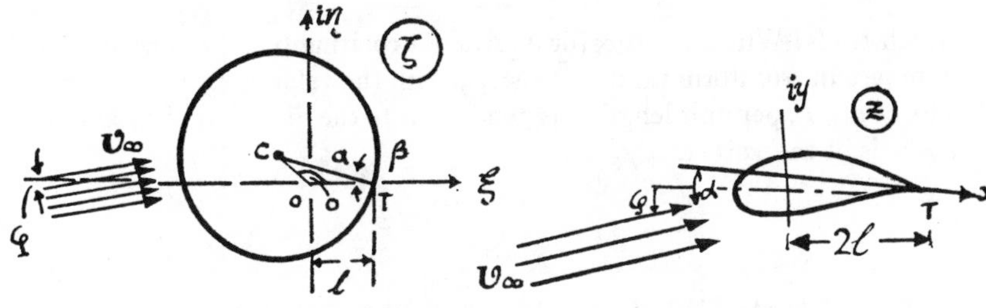

Fig. 67. Conformal transformation of a circle into an aerofoil

emerges. If, in addition, the larger circle's centre is placed on the oξ-axis, the transformed aerofoil will be symmetric. Then, varying the parameter of the cylinders, one obtains aerofoils of different configurations.

In all these transformations, the trailing edge of the aerofoils emerges as a sharp point of analytic singularity. This mathematical complication was resolved by the so-called Kutta-Zhukovsky Condition, or Zhukov-sky's Hypothesis, which states that the critical points of the transformation must correspond to the stagnation points of the flow in the plane of the circle.

These and other results were, however, the consequences of the more fundamental Kutta-Zhukovsky Theorem. The German mathematician Wilhelm Kutta (1867–1944) became interested in Fluidmechanics in connection with Otto Lilienthal's idea that curved surface wings were better than flat ones. Kutta wanted to work out a mathematical justification for the proposition, and arrived at the theorem in question.†

Zhukovsky arrived at the same result in a different way. Namely, back in 1890, he undertook a theoretical study of flight, and in 1898 attempted to extend his conclusions to what he termed 'winged propellers' – flapping wings. The construction of flow patterns around stationary wings was much easier than around flapping wings. But the great scientist was not prepared to be scared away by difficulties. He made numerous sketches on paper, prepared his mathematical tools, and – a very interesting *and* – played with strips of paper (size, for instance, 8 × 1 cm). You just drop such a piece of paper slightly above your face and you will see that it autorotates and refuses to fall vertically. Many contemporaries of Zhukovsky believed that this simple experiment played an almost decisive role in the discovery of the most fundamental law of aviation, the Zhukovsky Theorem, or the Kutta-Zhukovsky Theorem,

$$\boxed{L = \rho v_\infty \Gamma},$$

which reads: When a vortex (or equivalent rotating body) of circulation Γ moves in a uniform fluid of density ρ with the velocity v_∞, it produces a force $\rho v_\infty \Gamma$, per unit length, perpendicular to the direction of v_∞ and to the axis of the vortex.

Frederick Lanchester (1878-1946) and others

Helmholtz showed that if there is no initial vorticity in a fluid, then it can be created by friction or by a sharp edge of the body. In the latter case, a discontinuity in the flow may be formed between two fluid

† *Auftriebskräfte in strömenden Flussigkeiten.* Kutta, M. W., Illustrierte Aeronautische Mitteilungen, Nl. 6, 1902. *Aerodynamik.* von R. Fuchs, L. Hopf, Fr. Seewald, vol. I and II, Berlin: 1935.

streams meeting behind the body. This discontinuity can be considered as a continuous sheet of small vortices behind the trailing edge of a wing.

Kutta and Zhukovsky studied, however, ideal fluids, and assumed that the streams arrive at the trailing edge with one and the same velocity, therefore the problem of determination of the circulation Γ became a matter of straightforward mathematical operations. But it follows from their analysis – and experiments endorse this – that $v_l < v_u$ (Figure 68).

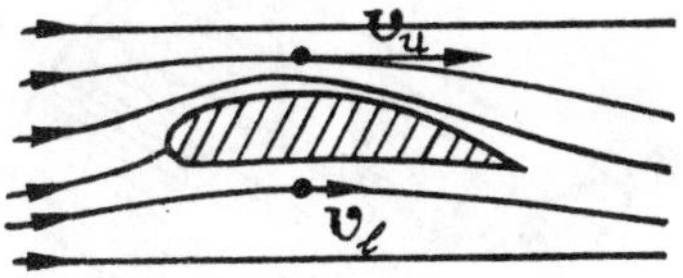

Fig. 68. Flow past an aerofoil

Suppose, then, $v_u = 3v_l$. If we subtract from both v_u and v_l the quantity $2v_l$ (say), we shall have: $v_u - 2v_l = 3v_l - 2v_l = + v_l$ on the upper side: $v_l - 2v_l = - v_l$ on the lower side. Performing similar operations for a number of other pairs of points, and then plotting the results with their signs, we have the picture shown in Figure 69 (arrows).

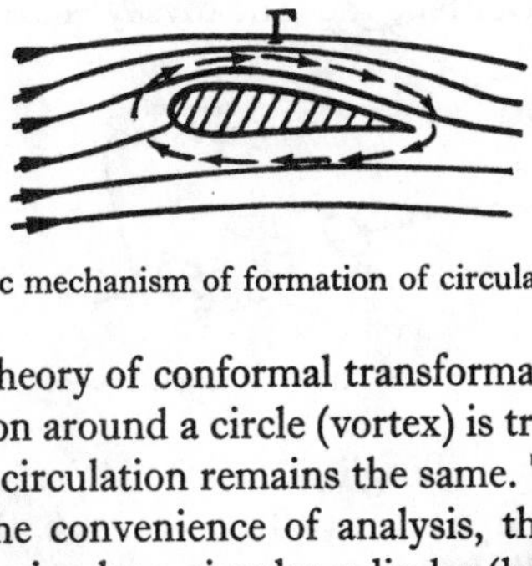

Fig. 69. The kinematic mechanism of formation of circulation around an aerofoil

The Zhukovsky theory of conformal transformation shows that when a flow with circulation around a circle (vortex) is transformed into a flow past an aerofoil, the circulation remains the same. Therefore Zhukovsky assumed that, for the convenience of analysis, the aerofoil can be replaced by a circle, a wing by a circular cylinder (long vortex). The latter is known as Zhukovsky's Bound Vortex – bound in the sense that it is imagined to be inside the wing, and confined to the wing (Figure 70). But no aircraft wing can ever be infinitely long, therefore the bound vortex, too, must have a finite length, or span: this contradicts Helmholtz's theorem, which demands that a vortex cannot end or begin in the fluid, it must end at the wall or form a closed loop or extend into infinity.

How can this difficulty be overcome? Let me say that I, personally, have no doubt in my mind that N. E. Zhukovsky must have been the first to suggest that his bound vortex twists at the tips of the wing and thus a П-shaped (or horseshoe) vortex system is formed (Figure 71). For, indeed, it was he who gave us the Zhukovsky Family of Aerofoils; it was he who worked out the basic mathematical methods of determination of the lifting force of a wing; it was he who delivered the first (in Russia) lectures on aircraft design and mechanics of flight; it was he who worked out the famous Vortex Theory of Propellers.

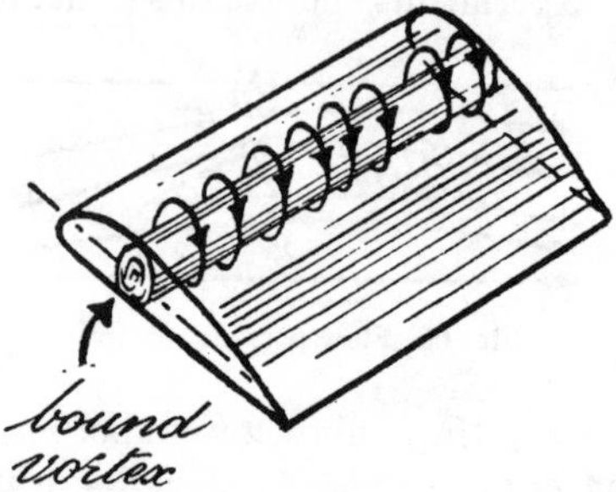

Fig. 70. The bound vortex

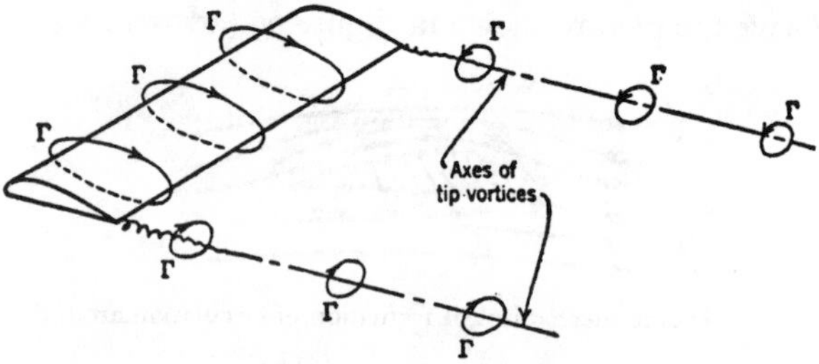

Fig. 71. The Π-shaped vortex

Which leads me to an English engineer, Frederick W. Lanchester (1878–1946), who, independently of anyone else, deduced from Helmholtz's theorem that the bound vortex does continue behind the tips of a wing of finite span, in the form of free vortices – free in the sense that they are not inside the wing – as shown in Figure 72. This was an outstanding contribution to the fluidmechanic theory of a wing. It not only clarified a great difficulty, but opened the gates to a whole series of additional problems and solutions.

It may be of interest to mention here that, in connection with a work called *The Aerodynamic Characteristics of a Wing with Moving Surface* (published by the Zhukovsky Academy of Aeronautics in 1942), I too, carried out in 1937–41 an extensive series of experiments on the mechanism of formation of the Lanchester free vortices, and this was the explanation produced (1938): since pressure under a wing (or a cylinder) is higher than above it, masses of the air tend to rush upwards through all the edges of the wing. But they cannot do so through the leading and trailing edges, because the oncoming flow carries them downstream.

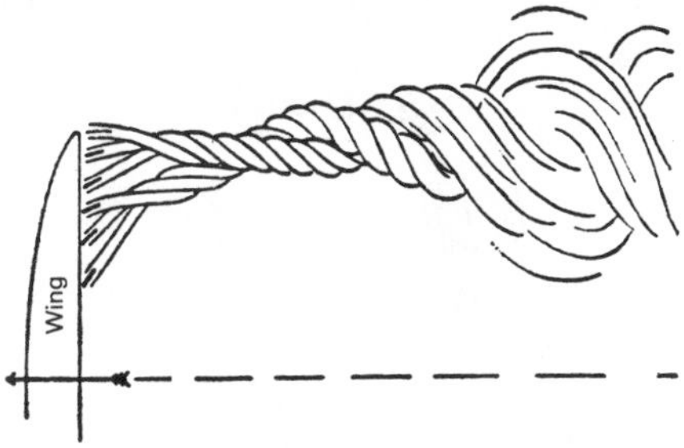

Fig. 72. Lanchester's 'free vortex'

The same happens to the masses of air trying to move upwards through the tips of the wing; but here we have something very different, that is, the masses are not only compelled to change their directions of motion, not only swept back and downstream, but also twisted together like the strands of a skein of wool. This is perfectly understandable, because these tip moustaches are nothing other than parts of one and the same bound vortex; the latter is located inside the wing and, theoretically speaking, replaces the wing, while the tip vortices are the continuations of the bound vortex beyond the wing, and extend into infinity, thus satisfying Helmholtz's condition. The kinematic mechanism of formation of the tip vortices is shown in Figures 73 and 74.

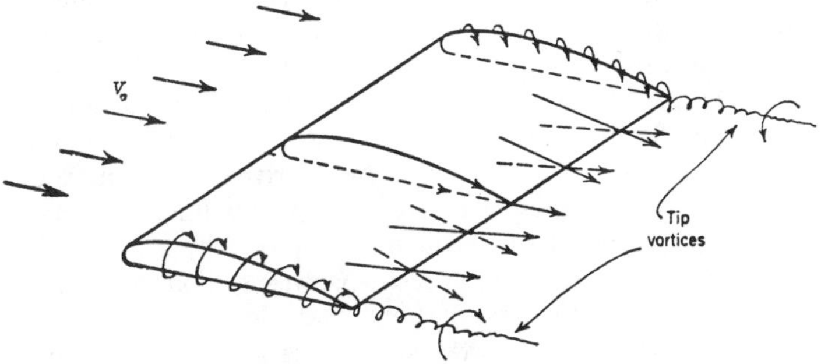

Fig. 73. The mechanism of formation of the П-shaped vortex

One doubts whether the word 'free' is the right word, since the vortices are not really free, but are caused and conditioned by the flow conditions at the tips, and are in this sense not detached from them. But whatever their name, the fact of their existence became known a long time ago, and can be shown both visually and quantitatively. Plate 3 shows the tip vortices and vortex sheet behind an airplane in flight, reproduced from Theodore von Karman's *Aerodynamics*. It is a photograph

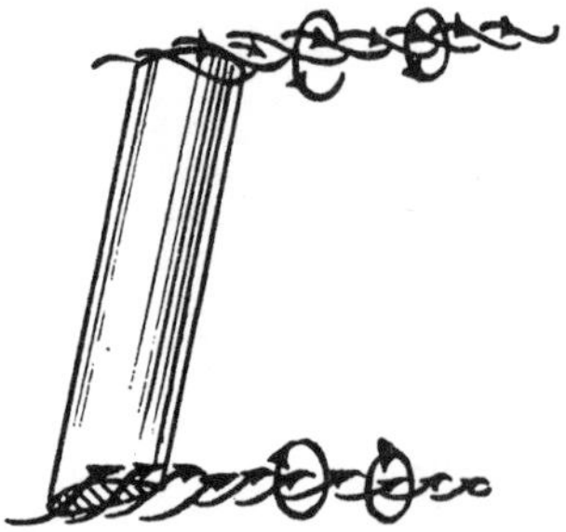

Fig. 74. Wing-tip vortex 'moustache'

of an airplane flying over a forest and emitting insecticide dust from its trailing edges. The free vortices can be seen very clearly, and immediately behind the whole wing in the span between the tip vortices there is a whole blanket of weaker vortices – a continuous vortex sheet.

The Prandtl-Lanchester Lifting Line Theory

I have shown the manner in which the basic scheme of the bound vortex and tip vortices was finally established: the next question was how to proceed from this scheme and from the Kutta-Zhukovsky theorem to the development of practical methods of aerodynamic evaluation of an aircraft wing of finite span? The problem was analysed by a number of theoreticians,† but, in the end, all of them arrived at one and the same method: the wing is replaced by a straight vortex line; the circulation about the wing associated with the lift is replaced by a vortex filament along the straight vortex line; at each spanwise station, the strength of the vortex is assumed to be proportional to the local intensity of the lift; the spanwise variation of vortex strength is assumed to result from superposition of a number of Π-shaped (horseshoe-shaped) vortices (Figure 75).

† *Applications of modern hydrodynamics to aeronautics*. L. Prandtl, NACA Rep. 116, 1921. *Coll. Works*. S. A. Chaplygin, Moscow: 1933 (and others).

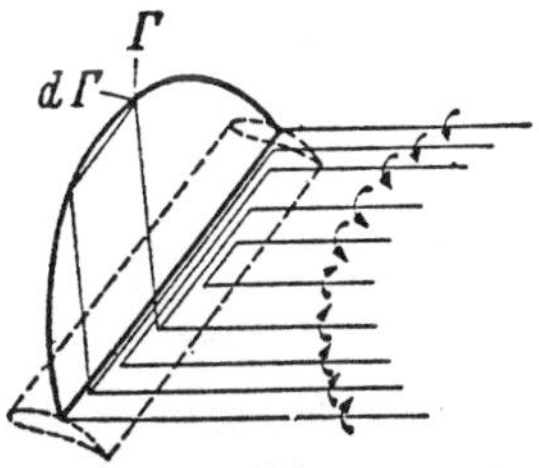

Fig. 75. The basic scheme of the theory of a wing of finite span

Obviously, the intensity of each vortex behind the trailing edge of a given wing depends on the distances from the wing tip or, which amounts to the same thing, from the tip vortex. This means that ε and $\triangle v_i$ (Figure 76), too, depend on the same distance (the only kind of

$$\alpha = \alpha' + \Delta\alpha_i.$$

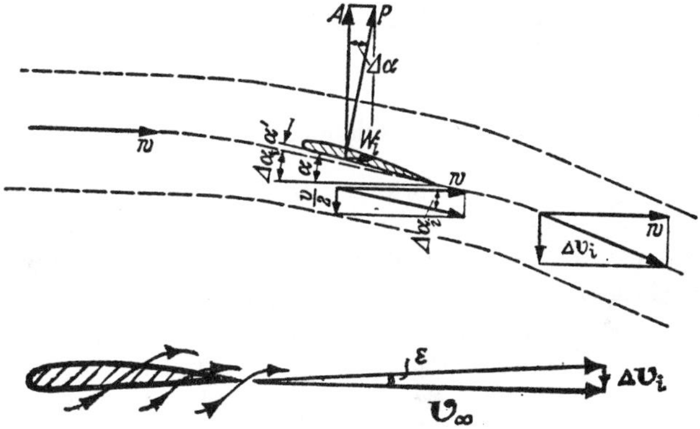

Fig. 76. The downwash angle and velocity

wing which produces constant $\triangle v_i$ along the span is one having an elliptical planform). It is very important to know the values of E and $\triangle v_i$ at each cross-section of the wing. The solution of this extremely complex problem was provided long before man knew anything about the downwash itself, by men who had nothing to do with Fluidmechanics. One of them was Felix Savart (1791–1841), a teacher in Paris, later the head of the physical cabinet at the College de France, the same man who showed that sound waves are propagated in water in the same way as in solids. The other was Jean Baptiste Biot (1774–1862), an outstanding French mathematician, physicist and astronomer. One of their many mathematical equations was adapted to the problem we are discussing, and is known as the Biot-Savart Law,

$$dv_i = \frac{\Gamma}{4\pi r^2} \sin \theta dl.$$

Now, since the downwash means simply that masses of air are being pushed downwards, i.e. imparting to them a certain downward momentum, there must be a force acting in the opposite (upward) direction, i.e. the downwash of a wing creates a lifting force. This was an important conclusion in several respects. First, to create and to maintain a downwash, a certain amount of kinetic energy (and, consequently, of work) is required. Secondly, as can be seen from the Biot-Savart formula, the amount of this energy varies along the span: the further away from the tips, the less the amount; which means, as was pointed out by Lanchester, that the greater the span (or, using aerodynamic terminology, the greater the aspect ratio) of the wing, the smaller the downwash and, consequently, the less the energy needed for the 'pushing down' of the masses of air, and hence the famous conclusion that a wing of an infinitely large aspect ratio has no downwash at all.

Ludwig Prandtl (1875–1953) of Germany worked along the same lines as Kutta, Zhukovsky, Lanchester and others. But he became convinced that the simple horseshoe vortex scheme did not always give results entirely satisfactory for practical purposes. He therefore introduced some additional ideas† which proved to be valuable. The Vortex Sheet I mentioned earlier is, really, the normal, natural consequence of the Zhukovsky Bound Vortex and of Lanchester's Free Vortices. But its kinematic details were worked out by a number of people, for example by S. A. Chaplygin,‡ Finsterwalder,§ Lanchester, Glauert, and others. Prandtl's work was accepted, however, as the satisfactory sum-total of all the contributions, mainly because of its mathematical simplicity and clarity, and became in due course known as the Prandtl-Lanchester Lifting Line Theory.

The essence of this theory can be described as follows. The velocity at any point of the space between the tip vortices is obtained from the vector sum of the component velocities arising from each of the three filaments of the horseshoe vortex. The lift distribution along the span is

† *Ergebnisse und Ziele der Göttinger Modellversuchanstalt.* L. Prandtl, Z.F.M., No. 1913. Also *Tragflügeltheorie I, II,* Göttinger Nachrichten: 1918.

‡ *Resul'taty teoretitcheskikh issledovanni dvizheniya aeroplanov* (1910), Coll. Works, vol. II, pp. 230–245, Moscow: 1948.

§ Z.F.M., No. 1, 2, 1910.

represented by a number of spanwise strips of constant lift per unit span, each having its corresponding horseshoe vortex. As the number of strips is increased, the actual lift distribution is represented more and more accurately, and the number of horseshoe vortices increases. In the limit the system becomes a sheet of trailing vortices. The influence of each bound and trailing vortex in the system contributes to the resultant velocity vector at any point in space, according to the Biot-Savart formula. The determination of the velocity field makes it possible to determine also the so-called induced angle, induced drag, and other associated phenomena.

One of the best books ever published on the aerodynamic theory of a wing of finite aspect ratio, which embraced all these methods and techniques, was called *Inducktivnoye soprotivleniye kryla*, the induced drag of a wing, Moscow, 1926, by B. N. Your'yev.

An important contribution to the theory in question was made by H. Glauert of Cambridge. In a book called *The elements of aerofoil and air-screw theory* (1926), he applied a Fourier series analysis to the solution of the problem of linearity of the lift curves, and developed methods for obtaining solutions for wings of any planform and twist. These methods were used in due course by Raymond Anderson, Robert Jones, V. V. Golubev (see his 'Teoriya kryla aeroplana konechnogo razmakha, TsAGI, Report No. 108, 1931), Barbara Fuchs, W. Richter, Betz, Max Munk and others, and so the problem of the wing of finite span was solved satisfactorily and finally.

Flettner's rudders

The history of Fluidmechanics would be pitifully incomplete without the story and history of Anton Flettner. From his early youth, he occupied himself with inventions of various kinds. No wonder he soon became associated with another great German inventor, Count Zeppelin. Flettner became known in his country and abroad in the summer of 1915, when a curious, clumsy vehicle made its appearance in Berlin: a tank which could move in any direction without a crew. The whole programme of the tank's manoeuvres was being directed from an unknown spot in the distance. The era of remote control had begun.

This success was so impressive that Flettner was soon afterwards appointed to the scientific staff of the Luftwaffe department. From then on, he was closely associated with many aerodynamic and hydrodynamic developments. First, he proposed to steer the large rudder surface of a ship by a secondary rudder. Its introduction and the understanding of the physical basis of its work were hampered by the fact that in technical circles, among most of the marine engineers, there existed an erroneous conception about the forces acting on the rudder. They thought that the rudder laid port or starboard would experience a current pressure, so to speak, on the side facing the current: the familiar Newtonian impact theory. He had to argue and to demonstrate that, in addition to the impact pressure on the face side, a relatively low pressure region existed on the opposite side, and that the effectiveness of the rudder was determined almost entirely by the difference of pressures in these two regions.

The still young aviation industry reacted differently. It was anxious to advance rapidly, it wanted new ideas and audacious solutions. Aircraft designers knew only too well that the power of air flows drove the old Egyptian windmills; that ordinary winds were capable of uprooting solid trees and destroying edifices; but that man was still very far from being able to understand or to use them fully. They knew that there is nothing new under the sun, but that the truth of nature lies hidden in mines and caverns for man to discover and to utilize. In short, aerodynamicists took Flettner's rudder idea so seriously, that the marine experts had to follow their example.

This is how Flettner's fluidmechanic ideas and inventions emerged. In utilizing the flow energy, we are almost exclusively dealing with surfaces which are subjected to a pressure by the current, he wrote.† Plate 4 shows how the current moves around a plate placed perpendicularly to its direction. One notices how the streamlines separate, and how, behind the plate, two vortex regions are created. Immediately in front of the plate a certain pressure (Standruck) is generated. High pressure in front, low pressure (suction) behind: hence, the resistance experienced by the plate, of which about two thirds is due to pressure and about one-third to suction.

In Plate 5, the same plate is inclined to the current. Conditions now have become unsymmetrical. The separation takes place at the leading edge of the plate. Everything is different. The centre of pressure is now near the leading edge. A change of angle of attack has changed the pattern of flow and shifted the centre of pressure. How could he, Flettner, exploit these facts and phenomena? To remain indifferent would be against his nature. The restless mind of an inventor backed up by a fairly good knowledge of the basic laws of Fluidmechanics wanted to explore, to experiment, to design, to test. In many cases, he sought his

† *Mein Weg Zum Rotor* (The Story of the Rotor). Anton Flettner, translated into English and published by F. O. Willhoff, New York: 1926.

answers in the history of science and technology. There are proofs in his writings that he studied the history of Fluidmechanics quite seriously. Since prehistoric times man has tried to utilize the forces produced by fluid flows, he wrote. But even the windmill, in most recent times, has not progressed beyond the dimensions of the old Dutch Windmill, while, on the other hand, the waterwheel has evolved into the modern turbine. The development of shipbuilding was rather slow, but on the other hand, the development of the aeroplane was very rapid. And so on, and so forth. Why such contrasts? Because in some fields man tried harder than in some other fields. And this means, therefore (continued Flettner), that there are good reasons for being optimistic: the treasures of the world of fluids will not disappoint the persistent explorer.

A fluid flowing past a body loses part of its energy, because the body configuration compels its particles to deviate from their free trajectories, and thus lose part of their energy of motion. This lost energy 'acts upon the body as flow resistance', according to Flettner, 'and to balance such a resistance, which is air resistance, the energy of an engine is needed'. It follows from here, he continued, that if a fluidmechanic design of any kind is to be economical, it must have a configuration producing as little disturbance to the particles of the fluid as possible.

Flettner points out, then, as an illustration, that the air resistance of a body depends not only on its size, but also on its configuration. 'Thus, it is quite possible that two airships, having exactly the same air displacement, may have different resistances at one and the same flow speed, which shows that the design of airships requires a special study aimed at the creation of streamlined shapes of minimum resistance, that is, shapes offering least disturbance to the flow.

That, however, is not all, as Flettner warned. The condition of the surface also plays an important role in the formation of the air resistance force; bodies with rough skin or corrugated outer surface offer greater resistance than bodies with smooth surfaces. 'It follows from here that the study of the nature of interaction between fluid flows and solid surfaces is very important'.

We may think today that there was nothing new in these and similar statements, but we should bear in mind that Flettner spoke before and during the First World War, when there were still no Nikuradse's experiments or even finalized theories of bodies of least fluidmechanic drag. For his time, he certainly was a prominent pioneer. But he was beating the already beaten drum when he declared that the fluidmechanic resistance of a body increases approximately as the square of the flow velocity, and this holds for medium velocities: this was already a universally accepted law. He was also wrong in asserting that the Rv^2 law was 'good' only up to about $v = 30$ miles per hour: it was 'good' up to very much higher speeds.

The reader will recall the Reynolds number. Flettner considered it an important parameter. In addition to the shape and the flow velocity, he wrote, quite rightly, this Reynolds ratio affects the resistance experienced by the body; it may happen, in some cases, that, in spite of the increase of the flow velocity, the fluidmechanic resistance suddenly drops.

As we have already indicated, bodies moving in fluids create waves. The principal difference between a 'watership' and 'airship' is, we read in Flettner's book, due to the fact that in the case of the former, in addition to the drag already discussed, the wave formation has to be taken into account. This wave formation alters the laws of the flow in an almost fundamental manner. For instance, the increase of the resistance with the velocity is entirely different from the ordinary $R \propto v^2$ law.

We can therefore appreciate that Flettner was more than an inventor: he was a mature aerodynamicist and hydrodynamicist. He cannot be treated too lightly and looked at as a blind empiricist, as just an inventor.

Let us recall that in order to cause or to stop rotation of an aeroplane about any of the three axes, and in order to keep it in straight level flight, or to execute various manoeuvres, it is necessary to have control surfaces movable with respect to the wings and to the fixed tail-planes. The control surface for producing or regulating yaw is the movable vertical surface, the rudder, attached to the (vertical) fin. The rudder is actuated by the feet of the pilot: pushing forward with the right foot on the rudder bar moves the rudder to the right and causes a right turn. Similarly, the pushing forward of the left foot causes a left turn.

The control surfaces for producing or regulating pitch are the horizontal tail surfaces. The movable part is the elevator or the 'flipper', and the fixed part is the stabilizer. The elevator is actuated by means of a control stick; pushing forward on the stick causes the nose of the aeroplane to go down; pulling back on it causes the nose to go up.

The control surfaces for producing or regulating roll are the ailerons attached to the wings. They, too, are actuated by the stick (on a small plane), or by a steering wheel (on a large plane); moving the stick (or turning the wheel) to the right causes the left aileron to go down and the right aileron to go up, producing a roll to the right, and vice versa.

All these actions require the pilot's muscular force, and this led Anton Flettner to the question: how can the forces (or, to be more precise, the moments of the forces) needed to push the rudder bar and the control stick be reduced?

The answers he gave were and remain extraordinarily important and clever. The so-called Flettner's Diagram (Figure 77) shows that the centre of pressure of an ordinary flat plate exposed to a fluid flow travels from the leading edge (at zero angle of attack) to its geometrical middle (at $\alpha = 90°$). This shifting of the centre of pressure should, perhaps, be called the *Afanzini Effect*, after an Italian hydrodynamicist, Afanzini, who established it during his experiments with flat plates in water.

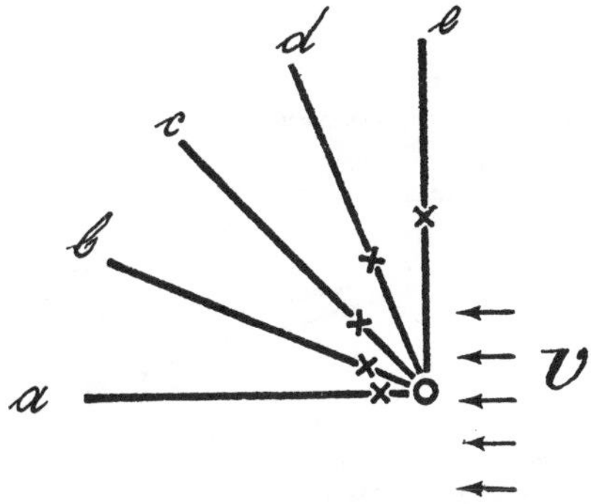

Fig. 77. Flettner's diagram of the Afanzini Effect

The fluidmechanic resistance of the plate is the greatest in position (e), in case (d) it is smaller, in case (c) still smaller, and so on.

This is what Flettner wrote:† According to the theory of the shifting of the centre of pressure, the balance – rudder – will always tend to assume a position where the turning axis and centre of pressure coincide (neglecting vortex formation). For all the practical purposes of shipbuilding, the rudder can be balanced for an angle of attack of 20°, because, at this angle, almost no force is required to hold it. Now, if it is desirable to turn the rudder from 20° into 30°, the centre of pressure will move in the downstream direction which, in the case of a large ship, would create a very considerable steering force. . . .

Flettner's idea was as follows: imagine a control surface of two parts: one allowed to turn about the hinge O, and the other about the hinge O'. The first of these develops a force R_1 and the second a force $R_2 \ll R_1$, so that its moment is small compared with the moment of R_1.

The resultant moment about the hinge O is equal to $R_1a - R_2b$, i.e. the stick moment (or the foot bar moment) is reduced by the amount of the second term. Thus, the addition of Flettner's *tab* to the rudder, to the elevator or to the ailerons, or to all of them, made it possible either to trim the aeroplane (hence the name 'trimmer') or to use the small

† *Mein Weg zum Rotor*, p. 25.

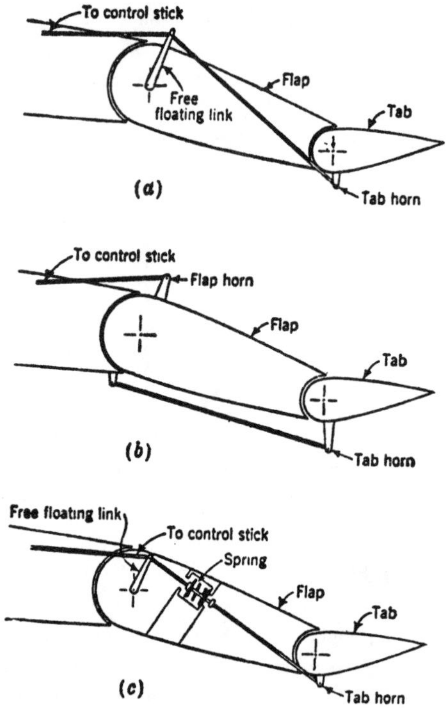

Fig. 78. Flettner's tab compensation

auxiliarly surface as a servo-device to enable the pilot to use less force in moving the stick and/or the foot bar (Figure 78).

This and some other devices, including the so called *Flettner's aerodynamic compensation*, have been used since 1920–25 both in the aeroplane and in all kinds of boats (Figure 79), and will always remind us that

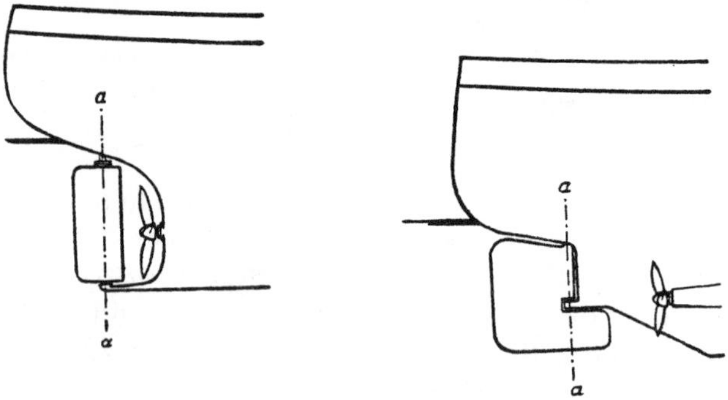

Fig. 79. Flettner's fluidmechanic compensations for boats

Anton Flettner occupies an outstanding position in the list of great men of Fluidmechanics.

Just one more remark before proceeding to Flettner's sensational invention. We all know that the sail-ship has serious disadvantages. The efficiency of the canvas sails is low, their control is a very troublesome job. No wonder they attracted Flettner's attention.

From 1920 until 1922, he lived in The Hague, Holland. 'By reason of the proximity of the ocean and the low elevation of the land, a wind of medium strength is blowing here almost continually', he writes.† 'The numerous windmills, large and small, are therefore running practically all the time. . . . Thus, I am a continuous witness of the energy of the winds. . . . I also have had the experience of voyages in stormy winds. I have taken part in the incessant, tense struggle of the sailors against the ever menacing forces of nature. . . . It is natural, therefore, that during my stay in Holland I have been working on the idea of revolutionizing the sail-ship. . . . And so I have come to the idea of *replacing the canvas sail by a metal sail. . . .*'

Why was this idea dropped, one wonders? I, for one, think that it deserves renewed study, because its potential is far from being insignificant, especially in view of the new achievements in sheet metal production.

Flettner's rotorship

Kutta's and Zhukovsky's theorem was and remains the foundation stone of aviation and of many problems in hydrodynamics; and the circulation, which we have already discussed, was and remains the heart of the theorem. From about 1922, Flettner's mind was occupied, therefore, with the question which should be prominent in the mind of every aeronautical student too: what is the 'Magnus Effect'?

'One morning', Flettner writes, 'while lying in the sand, I tried to explain to my wife the meaning of the Magnus Effect. I attempted to explain its peculiar action by rolling small grains of sand from a little sand hill, and making them at the same time pass my fist which I turned in the moving sand. One could clearly see that on that side of the fist where the direction of its rotation coincided with the direction of the moving sand, the grains moved very much faster; while on the opposite

† *Mein Weg zum Rotor*, p. 60. *Le rotor, instrument a progres en marge du 'Flettner-Rotor'*. B. A. Chait, Impr. Anvers-Bourse, Antwerp: 1925.

side they came to a standstill. . . . During the following night I could not sleep a wink. In those few hours I fought a battle with myself which later for months I had to fight with all the experts. . . . I was asking myself whether I should take it upon myself to introduce alongside my rudder system also the rotating sail, which seemed to be really revolutionary. My decision was rendered more difficult by the fact that already agreements had been concluded with the Germaniawerft to put the idea of the metal sail into practice. . . .'

This was, then, the beginning of the story of the Rotor ship. But I feel I must remind the reader once again that the Magnus Effect was, in fact, established by Benjamin Robins and Leonhard Euler. By this I certainly do not imply that there are no grounds for calling the effect after Magnus, but I mean to say that there are good reasons for not ignoring Euler and especially Robins.

Here are the facts. In 1794, the Berlin Academy of Sciences offered a prize for solving the problem of deviation of artillery shells (balls) from their theoretical trajectories. But it was 58 years later that Gustav Magnus, Professor of Physics at the University of Berlin, one of the teachers of Helmholtz, succeeded in giving the answer.† He constructed a brass cylinder which could spin in two conical bearings. An air flow created by a blower was directed at the spinning cylinder. As soon as the cylinder started revolving, he noted a lateral deviation towards the side where the wind and the rotating surface had the same direction: a fact whose reasons we have already discussed. Tennis players, too, know that if the ball is hit by the racket at an angle or 'cut', it flies *and* rotates; therefore on one of its sides the direction of its rotation is the same as the direction of its forward motion, and on the other side they are opposite; the flow becomes assymetric (Figure 80), and a Kutta-Zhukovsky lifting force is created, which makes the ball fly along a curved line.

Lord Rayleigh called this an 'irregular flight', and tried to explain its reasons.‡ Much later on – in 1912 – a French professor, Lafay, published an article under the title *Contribution experiméntale à l'aérodynamique du cylindre et à l'étude du phénomène de Magnus*, which contains no reference to Benjamin Robins or Leonhard Euler, but contains an elaboration of his own experiments in the Ecole Polytechnique and in the Ecole d'Aviation Militaire de Vincennes, in 1910–11 (Figure 81).§ From today's point of view, the experimental techniques were primitive. Nevertheless,

† *Uber die Abweichung der Geschosse und eine auffallende Erscheinung by rotierenden Körpern.* G. Magnus, Berlin: 1853.

‡ *On the irregular flight of a tennis ball.* Messenger of Mathematics, July 14, 1877: London.

§ *Prévision de l'action d'un vent dont la direction varie rapidement, application à l'effet Katzmayr et à l'autorotation.* A. Lafay.

they showed that the aerodynamic lift of a rotating cylinder was several times greater than that of an equivalent wing of the usual type.

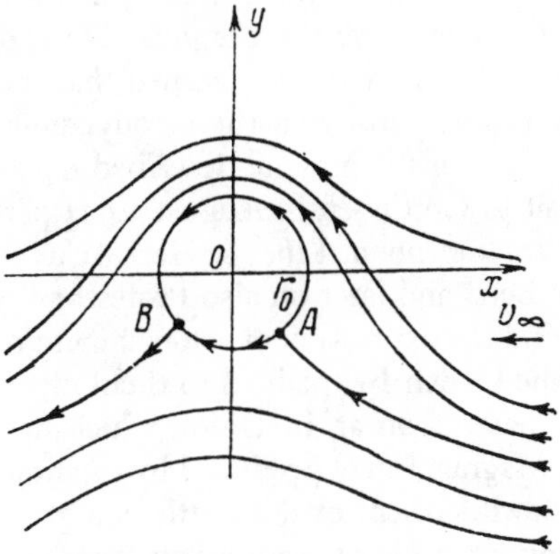

Fig. 80. Asymmetric flow past a rotating cylinder and/or ball

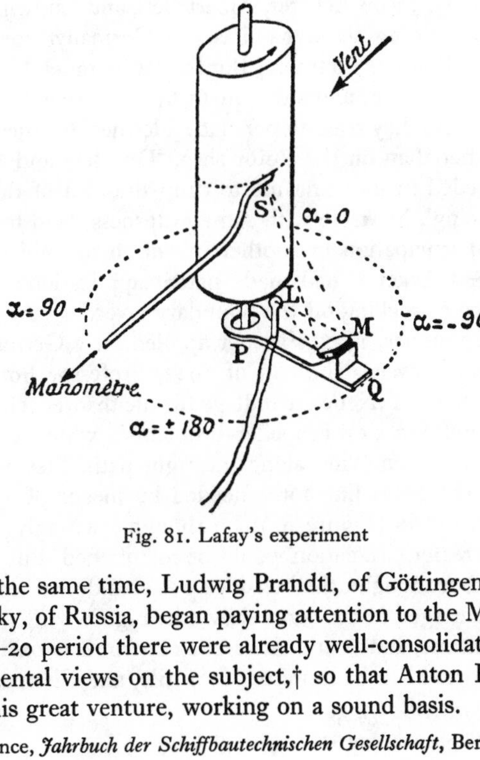

Fig. 81. Lafay's experiment

At about the same time, Ludwig Prandtl, of Göttingen, and Dimitri Riabouchinsky, of Russia, began paying attention to the Magnus Effect. By the 1918–20 period there were already well-consolidated theoretical and experimental views on the subject,† so that Anton Flettner could embark on his great venture, working on a sound basis.

† See, for instance, *Jahrbuch der Schiffbautechnischen Gesellschaft*, Berlin: 1918.

Flettner was now the leading champion and advocate of the practical uses of the Magnus Effect (he never called it the Robins' Effect, or the Robins-Euler Effect). Towards the end of 1922, the idea of the Rotor-ship, i.e. of a ship propelled by the Magnus Effect, occupied his mind more than anything else. In 1923, he learned that the Göttingen Institute of Aerodynamics was studying the aerodynamic characteristics of rotating cylinders, and this is how he described his experiences:†

'I recall my visit to Göttingen in August, 1923 particularly well. On my arrival in the late afternoon, I met in Herr Ackeret's office first Herr Ackeret and Herr Betz, and later on also Professor Prandtl. Even today I can see in my mind the expression of astonishment on their faces; they were truly astonished when I explained to them my plans and said that I had applied for patents on an invention which had as its object the application of the Magnus Effect produced by rotating cylinders for ship propulsion. The bewilderment of the gentlemen grew intensely when I told them about my intention of carrying out tests on such a ship at the Germania Werft.'

Needless to say, Jacob Ackeret, Albert Betz and Ludwig Prandtl, the three most outstanding aerodynamicists of Germany, knew at least as much about the Rotor as Flettner.‡ But they were much less enthusiastic about its practical value; and subsequent history proved that they were right. Accordingly, they tried to persuade Flettner to concentrate on his metal sails rather than on the Rotor ship. 'Dr. Betz and Herr Ackeret actually succeeded in inducing me to drop the idea of the Rotor ship for the time being', he wrote with some bitterness, 'and to consider the possibility of participation in another invention for which they themselves, Betz and Ackeret, had made patent application'. This was, of course, the idea of sucking off the boundary layer.

On 16th September, 1922, Flettner applied for a German patent for the Rotor ship. Towards the end of 1922, Professor Foettinger published an article called 'Neue Grundlage für die theoretische und experimentalle Behandlung des Propeller-Problems', which convinced the inventor that he was moving along the right path. Flettner's first idea was to create the propelling force needed by means of a belt moving around two cylinders (Figure 82): he thought (wrongly) that, in this way, a much greater circulation would be maintained. But, after several months of further thought, the belt-idea was abandoned, and the Rotor-ship concept emerged in the form shown in Figure 83.

In the meantime, Betz's and Ackeret's experiments on the boundary layer sucking proved to be more or less futile, and Flettner's invention

† *Mein Weg zum Rotor*, pp. 67–68.
‡ *Zeitschrift für Flugtechnik und Motorluftschiffahrt*, 3, 1925.

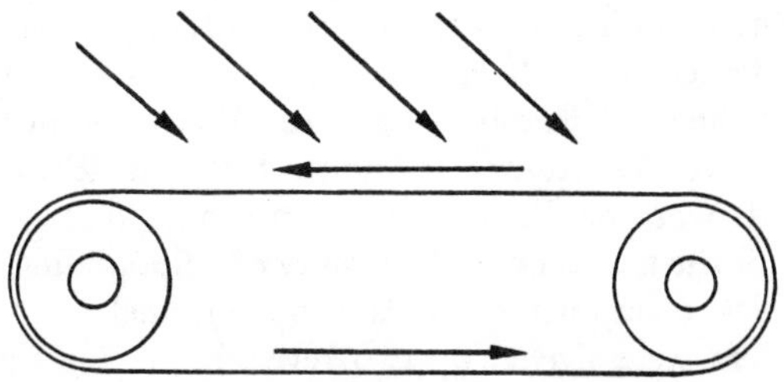

Fig. 82. Flettner's belt driven by two cylinders

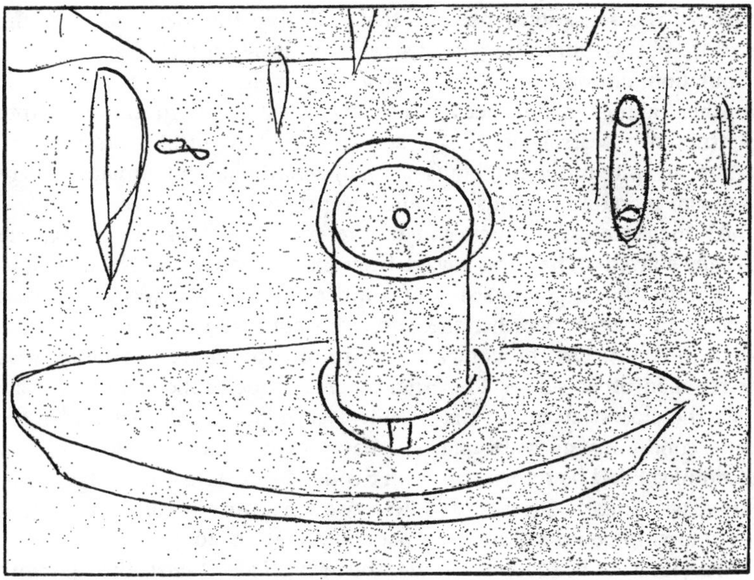

Fig. 83. Flettner's original scheme for a rotorship

began receiving more favourable attention. At long last, his models found their way into the Göttingen wind-tunnels (Plate 7). The experiments proved beyond doubt what was already known: that a rotating cylinder exposed to a wind creates on one side a region of low pressures and a region of higher pressures on the opposite side. The force arising from these pressure differences was the 'Magnus Effect', the aerodynamic thrust needed to propel a ship.

Anton Flettner was now ready to proceed to the next stage. Early in 1924, he invited representatives of the Germania Werft to Göttingen, where he delivered a lecture on the Rotorship. A number of technical objections was raised, but (note well!) Betz, Ackeret and especially

Prandtl were now on his side; with their support, he got the idea through the doors of basic approval. He first built an experimental rotorboat (Plate 8), and then, by October, 1924, the Werft completed the construction of a large two-rotor ship under the name *Buckau* (Plate 6). And they both worked! *Buckau* moved not only forwards, but even backwards! For the first time in the history of shipbuilding, here it was, in the Baltic Sea, a ship without sails or huge funnels.

The cylinders were driven by two reversible d.c. shunt motors of 11 kW, 220 volts and 750 r.p.m. The rotors themselves were made of thin sheet steel of $\frac{3}{64}$ inch thickness. As soon as the official trials commenced, the sensational news of the Rotorship thundered all over the world. 'It was interesting to observe how in almost all classes of the population, not only among the scientists and engineers, but in all kinds of other callings, people occupied themselves with my invention', wrote Flettner two years later. 'I saw and still see how in all classes of the nation, in spite of the present (i.e. 1926) conditions (in Germany), there exists an incredible enthusiasm and receptivity for new ideas.'

At the successful conclusion of the trials, in February, 1925, the *Buckau* started on her first voyage from Danzig across the North Sea, to Scotland. The cargo consisted of lumber. The voyage was a complete success. In stormy weather the rotors did not give the slightest trouble for a moment. On the contrary, the rotorship behaved in heavy seas better than other ships! The English press which followed the voyage with the closest attention was unanimous in conceding that under the same conditions no sailship could have done what the rotorship had accomplished. The return trip from Scotland to Cuxhaven, with a heavy cargo, was equally successful. These two trips proved conclusively that the rotorship could sail into the wind at about 20° to 30°.

On 31st March, 1926, now under the name *Baden Baden*, the rotorship sailed to New York, via South America. Then there were many other trips.

But . . . every stick in the world has two ends. However brilliant in its basic idea, no technological newcomer has ever been free from shortcomings. And every innovation has its opponents. So, the great fluidmechanic invention became the victim of its own shortcomings. The fact that the rotors could not produce a Magnus Effect in calm, windless weather, was used as anti-rotorship poison. . . .

Flettner's invention remains wounded, but is not dead. It imparted to Fluidmechanics a momentum of so great a magnitude that the rotor made its triumphant way into every laboratory of aerodynamics and hydrodynamics in the world, and into the mind of every aero- and hydrodynamicist.

Flettner's rotor windmill

The history of the use of windpower extends very far back. Apparently, there were windmills – as distinct from watermills – in China and Japan thousands of years ago. Even Hammurabi reports about windmotors which were intended to furnish power for the great irrigation systems of Babylon. In Egypt old mills still exist which are supposed to have been erected two thousand years ago. The Arabian explorer Istachri (134 AD) reports the use of windmills in the Persian province of Segistan. The Arabian scientist Dismaschgi (1271 AD) gives us a detailed description with illustrations of these old Persian windmills (A. Flettner).

This Persian mill had the windwheel, a kind of turbine mounted in the lower part of a structure whose walls were provided with wind inlet and outlet openings. The wheel turned on a shaft supported by a point from below. The wind entered the motor chamber through the openings on the side and blew into the bellying sails. The mill stones were mounted in the middle storey, while the top storey contained the hopper.

The Germans began using windmills in the year 833 AD, the French in 1105, the English in the twelfth century. From then on, man has tried to exploit the power of winds in various ways, but not always successfully. The first Dutch windmill for grinding grain is reported to have been erected in Holland in 1439. Throughout the next century or so, similar mills appeared in many countries, and so air flows began serving man in yet another way.

But the way was, and remains, both narrow and short. We must admit that this is one of those fields of human effort where man's achievements are still infinitely small compared with the potential yet untouched. Incessantly the sun, by its heat rays, sets in motion the enormous masses of the atmosphere, untold billions of horsepower are in the air in the form of winds, but all man has managed to create for their use is the primitive windmill. . . .

Yes, one of the oldest man-made machines, the windmill, is still very primitive. As Anton Flettner complained:

'The first problem which applies to windmills of any size, is the difficulty of its regulation. It is most difficult to calculate the wind force acting on the windmotor. The energy for which a windmill must be designed often fluctuates between several thousands, because the wind itself changes from a gentle breath to hurricanes which uproot giant trees. This great divergency between the forces to be calculated for the design of a windmill resulted in such uncertainty that the modern engineer, used to exact calculations, did not occupy himself with the problem of the wind power plant.'

How could these shortcomings be overcome? There were in the past many – we could even say very many – new ideas. But none of them was as revolutionary as Flettner's suggestion that the usual windmill wing (blades) be replaced by rotating cylinders. In 1923, a syndicate was formed consisting of the Deutsche Maschinenfabrik A.G., Duisburg (DEMAG), Mannesmann, and Bergische Stahlindustrie Remscheid, under the auspices of the Deutsche Bank, the object being to carry out, jointly with the inventor, experiments on the rotor windpower plant, and to look after its subseque industrial-commercial aspects. In 1925, this syndicate organized the Flettner Windturbine Company. Plate 9 shows a photograph of the first rotor windmill built and actually tested by the DEMAG.

The diameter of the windwheel was 66 ft, the rotors were about 16·5 ft long, tapered, with a diameter of 35·5 ft on the outer end, and 28·5 on the inside end. The tests showed that the rotor was eminently suited to replace the traditional windmill wings and sails. One of the advantages of the new machine was that even severe squalls and gusts of wind did not strike a wing or sail, but relatively small cylindrical drums, which had but one-tenth of the aerodynamic resistance offered by a sail of the same power output. At that there was no danger of 'running away' and reaching dangerous speeds, since the rotation of the cylinders could be easily controlled. But, as in the case of the rotorship, it had its limitations. Worse than that, the aerodynamic limitations of the windmill rotor were greater than those of the rotorship. In the first place, the Magnus Effect of a rotor depends on the ratio of the rotational velocity u to the wind velocity v, which could not be maintained in stormy winds. Secondly, when the main windwheel is revolving, the outer parts of its cylinders move through the air at much greater velocities than their inner parts, which complicates the aerodynamic characteristics very significantly. To overcome this, a drastic re-design was needed; but this reached beyond the borders of commercial reality.

At any rate, the fact is that Flettner's rotor windmill, like his rotorship, never got beyond the stage of a sudden flush of interest from all over the world, but man failed once again in his endeavour to exploit the energy of the winds.

Autorotating bodies

The purpose of this book is three-fold: to give an outline of the history of Fluidmechanics; to promote the re-examination and re-evaluation of some of the great but abandoned inventions and ideas like Flettner's rotorship and rotor windmill; and to show, by specific examples, that the history of science and technology is far from being free from misconceptions and the tyrannical influences of certain dogmas, which tend to repeat themselves decade after decade. Look into your textbooks and you will see striking examples: Bernoulli's Equation (Lagrange never mentioned), 'Magnus Effect' (Robins never mentioned), and so on.

The same can be said about the history of autorotating bodies, which constitute one of the more interesting branches of aerodynamics. We still do not know who first discovered them. But various people, at various times, have made direct or indirect claims. 'Once my inventions had shown that the 'Magnus Effect' some day was destined to play a significant role in engineering', wrote Flettner, 'after I had succeeded in bringing to the attention of the astonished world the significance of my discoveries, there began practically everywhere an intense activity in the new field, numerous patents were applied for; but I had already provided for these possibilities in my first patent application. . . .'

Careful examination of this somewhat pompous declaration reveals that it also includes autorotating bodies. But truth shows best being naked, and never grows old. I do not know whether Anton Flettner was familiar with the contributions of James Clerk Maxwell (1831–79) to knowledge, but I feel that he, Maxwell, would disagree with him rather sharply. We all know him as an outstanding Scotsman who translated the theories of Michael Faraday (1791–1867) into the language of mathematics, as the author of a theory according to which the energy of the electromagnetic field resides in the dielectric as well as in the conductors, as the creator of electromagnetic theories and theorems, and as the writer of the epoch-making book *Treaties on Electricity and Magnetism* (1873). But he was also an aerodynamicist, at least on a small scale. It was he who gave birth to the theory of autorotating bodies, long before Flettner and even Zhukovsky, and this is how he treated the problem.

Suppose that an autorotating lamina *A* (Figure 84) falls vertically. When it occupies the position *a*, the drag is much less than when it assumes the position *c*. Therefore when the lamina arrives at *b*, the relative wind v_1, consequently, the resultant force *R*, is greater than the relative wind v_2 (and, consequently, the resultant force *R*), in position *d*. On examining all the positions taken by the lamina, one sees easily that the body can, and does, continue its fall with autorotation, and that the resultant force *R* pulls the lamina to the left.

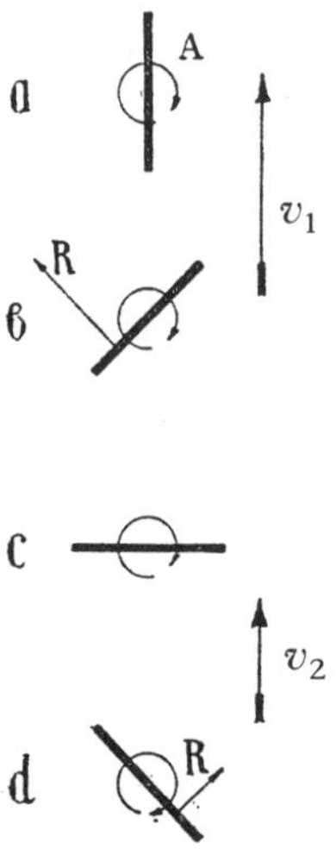

Fig. 84. Maxwell's scheme of autorotation

Here we have very clear statements: an autorotating body produces a Magnus Effect; it cannot fall vertically, it must move along a curved trajectory. We thus see that not only Robins, Euler, Zhukovsky, and Magnus, but also Maxwell could, perhaps, claim to be the author of the Magnus Effect; and how can anyone claim that aerodynamic autorotation was discovered only in connection with the Rotorship?

D. P. Riabouchinsky analysed Maxwell's theorem a step further. When the strip *A* (Figure 85), after having occupied position *a*, takes position *b*, the curvature of the streamlines of the relative motion and, consequently, the suction effect in the vicinity of the leading edge, driven by the wind, are stronger than when lamina *A* takes the position *d* after having passed through the position *c*. The consequence of this explanation is that in order to increase the speed of the autorotation, it is necessary to accentuate the suction effect behind the edge driven by the wind, for instance, by cutting its edge at an angle.

In accordance with this consideration the three plates *A*, *B*, *C*, the

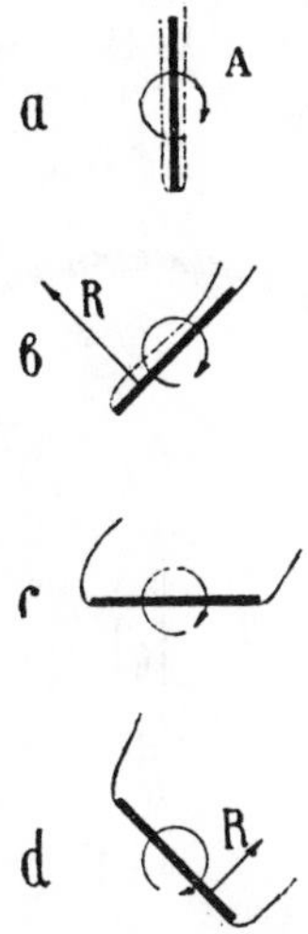

Fig. 85. Riabouchinsky's explanation of the causes of autorotation

section and rotation direction of which in respect of the stream velocity are indicated on Figure 86, must rotate at different speeds. Plate *A* must rotate at higher speed than plate *B*, and plate *B* faster than plate *C* (Riabouchinsky's Theorem).†

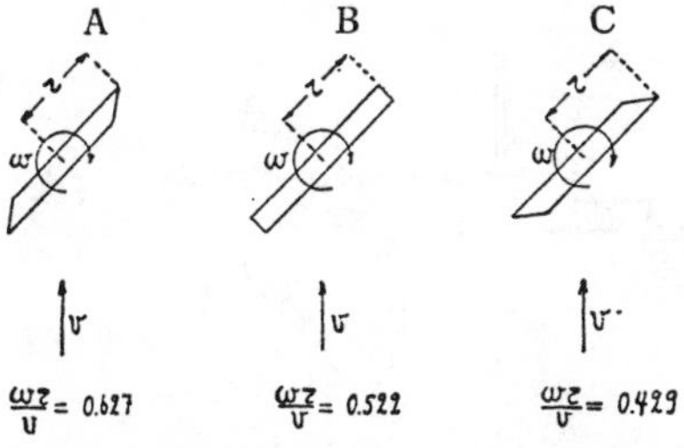

Fig. 86. Riabouchinsky's theorem of autorotation

Figure 87 shows the arrangement used by Riabouchinsky for the study of the autorotation of these plates in the wind tunnel. The autorotation coefficients, i.e. the ratio of the edge velocity $u = \omega R$ to the stream velocity V, of plates A, B, C, were found by him to be respectively 0·627, 0·522, 0·729. In autorotation of this type, i.e. around an axis orthogonal to the relative stream, one has always $\omega R/V < 1$. Plate 10 shows photographs of smoke filament lines in the neighbourhood of a plate set in autorotation in an airstream. The arrangement used is shown in Figure

† *Thirty Years of Theoretical and Experimental Research in Fluid Mechanics*. Dimitri P. Riabouchinsky, Royal Aeronautical Society, London: 1935; also: *Etude sur l'hypersustentation et la diminution de la résistance à l'avancement*. Dimitri Riabouchinsky, SDIT, Paris: 1957.

88. These photographs (obtained in 1930) concord with Riabouchinsky's explanation of the aerodynamic reasons for the autorotation of flat plates.

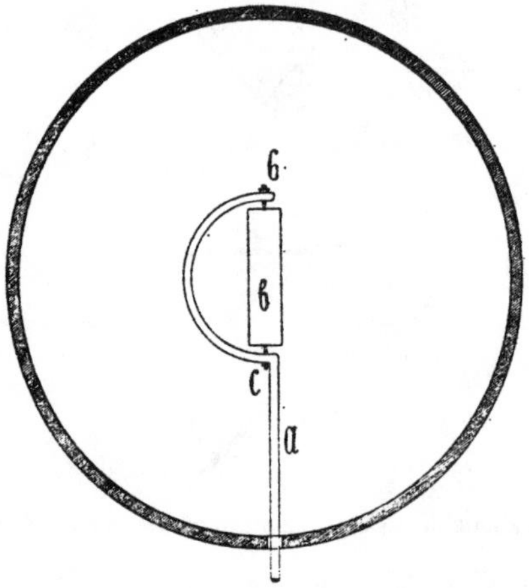

Fig. 87. The arrangement used by Riabouchinsky for the study of autorotation in a wind-tunnel

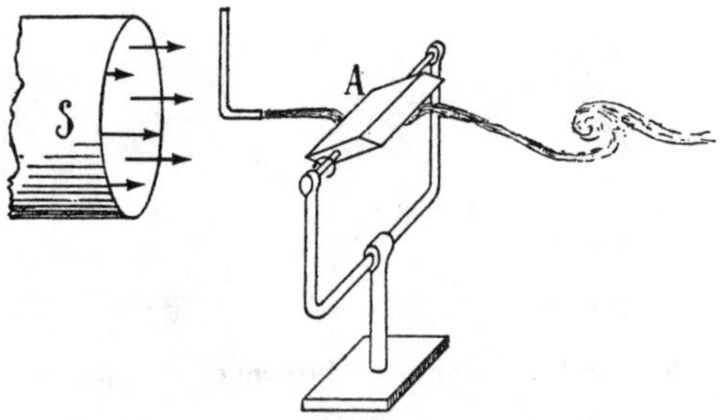

Fig. 88. Riabouchinsky's method of visualisation of flows past autorotating bodies

Otto Lilienthal discovered that, for certain angles of incidence, the resultant aerodynamic force on a curved aerofoil can be not only perpendicular to the chord, but even deviate towards the leading edge of the aerofoil. An interesting application of this fact was made by a Danish scientist, Paul la Cour,† who constructed a windmill capable of rotating

† *Forsog med smoc Mollenmodellen*, p. 62. Ingenioren: 1899.

in both directions; if the wings of the mill were flat, such a deviation of the aerodynamic force and, consequently, such indifference to the direction of rotation would be impossible.†

In 1905, at the 4th International Congress of Aerostatics, in St Petersburg, a certain Patrick Alexander demonstrated a small rotor capable of rotating indifferently in both directions. Some weeks later, Riabouchinsky, too, carried out a series of experiments, with special attention to the pressure distribution in the vicinity of the leading edge, and worked out an explanation of the property of autorotation of a flat plate (Figure 89).

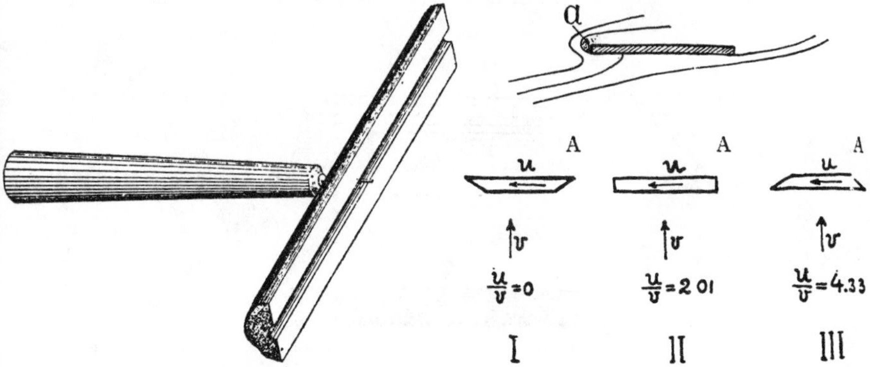

Fig. 89. Riabouchinsky's method of investigation of Patrick Alexander's discovery

So, the basic concepts of autorotation were established. But what were the aerodynamic characteristics of such bodies, and could they be useful to man?

In the Koutchino Institute of Aerodynamics near Moscow, D. P. Riabouchinsky carried out free-stream and wind-tunnel tests of one-, two-, three- and four-bladed rotors (Figure 90), and the results are given in Figure 91.‡ The symbols used are: M = density, N = number of revolutions per unit time, R = rotor radius, $S = 2Rh$, h = the length of the rotor; P_1, P_2 and P_3 = lift coefficient curves; L_1, L_2 and L_3 = drag coefficient curves, the subscripts indicating numbers of blades.

Much later on, probably in 1922–3, a Finnish engineer, Commander Sigurd J. Savonius, suggested a new type of autorotating body,§ which, I understand, he initially intended to call 'TSPM' (The Savonius Perpetuum Mobile); in actual fact, however, it became known as

† *Die Windkraft*. Leipzig: 1905.
‡ *Bulletin de l'Institut Aerodynamique de Koutchino*, No. 3, p. 34. December: 1909.
§ *El Flugerrotor*. Iberica, Ano 14, Num. 708, 31.12.1927.

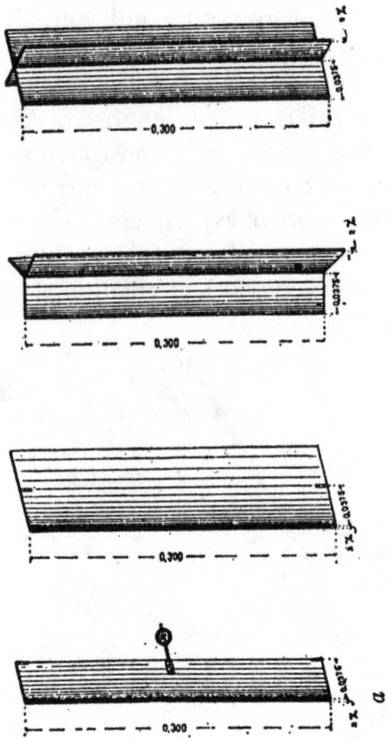

Fig. 90. Riabouchinsky's one-, two-, three- and four-bladed rotors

'TSWR' (The Savonius Wing-Rotor), or simply wingrotor: two half-cylinders shifted in respect to each other (Figure 92). The device found its way into many wind-tunnels in the world, including those in the USSR, and a number of projects for its use were proposed, for instance, the use of the *wingrotor* as a windpower motor to drive small windmills.

The main advantages of the wingrotor over all the other autorotating bodies were the steadiness of its autorotation and relative insensitivity to (small) changes in the speed of the wind. These properties made Savonius' invention an almost ideal driver of ventilators in railway wagons, restaurant-cars and ordinary van-vehicles. On the other hand, there were all over the world loud speculations about possible uses of the wingrotor as a water pump drive, and even as an effective aircraft wing.

It is understandable, therefore, that we in the USSR were not indifferent to this *perpetuum mobile*. It was studied in the Central Aero- and Hydrodynamic Institute from about 1927–8 onwards. Ten years later, I myself was instructed to carry out in the large wind-tunnel of the Zhukovsky Academy extensive investigations on rotors, autorotating

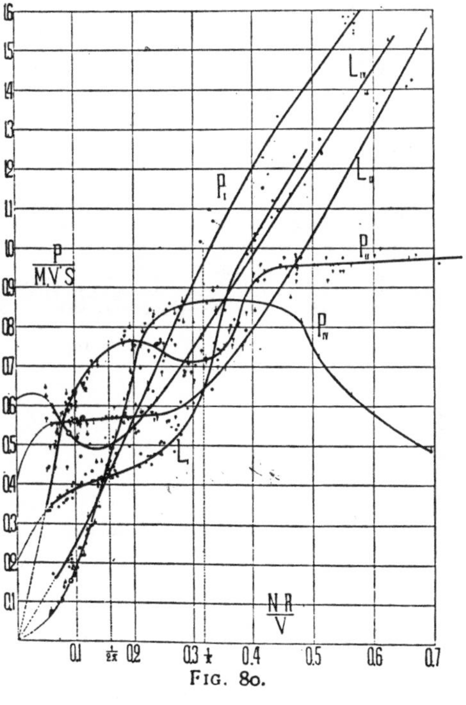

Fig. 91. Riabouchinsky's experimental curves of autorotation

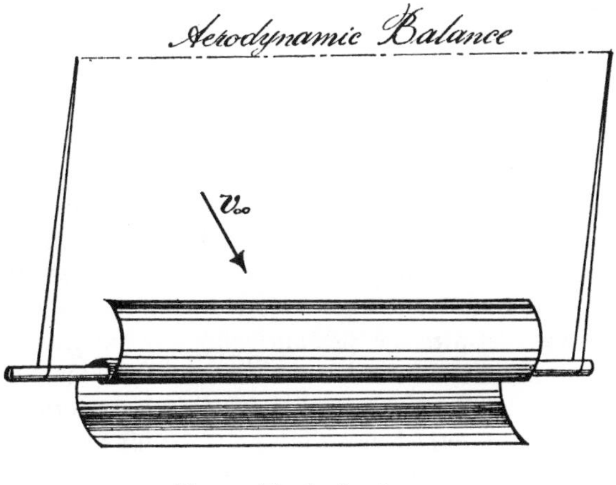

Fig. 92. The basic wingrotor

bodies and – a very interesting 'and' – with *combined* rotors (Figure 93). The essential aerodynamic characteristics of the wingrotor emerged in the form shown in Figure 94 (tip discs increase the value of C_{Lmax} noticeably); but the characteristics of the combined rotors were roughly the same as for ordinary rotors.

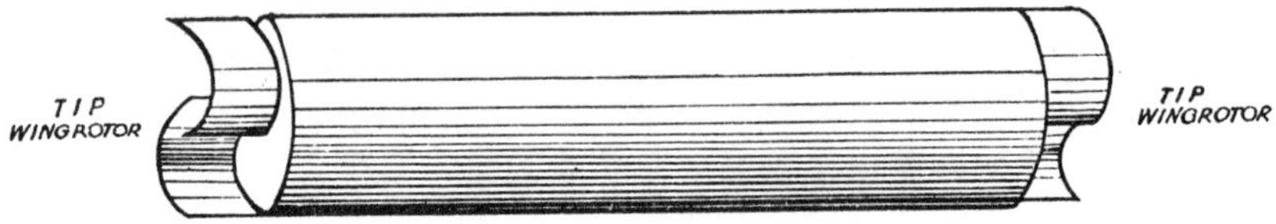

Fig. 93. Tokaty's combined rotors

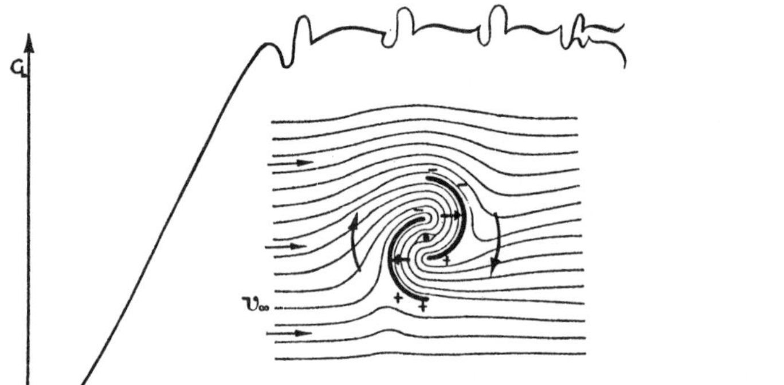

Fig. 94. The aerodynamic characteristics of a wingrotor

I suppose the average reader is familiar with the toy aircraft with rotating plates – nearly wingrotors – instead of wings: this model was proposed in 1901 by a German called Köppen. An old idea, but we were certainly interested in it. In 1939, I carried out wind-tunnel and free-wind tests of a model with the Savonius wingrotors instead of the usual wings. I then proceeded to similar experiments of the same model with the combined rotor instead of wings. Both models performed well, the idea seemed to have worked, but the amount of lift produced by them was far from being anything like adequate, therefore they received no further attention.

The late Theodore von Karman once remarked that there were in the history and philosophy of science and technology 'written truths and verbal truths'. Now, according to the 'written truths', the experimental aerodynamic characteristics of rotating cylinders were obtained first by Jacob Ackeret in about 1924. One can prove this by numerous publications.†

But let us keep in mind that the progress of knowledge is often hampered by the tyrannical influence of certain printed 'truths' that at the end of the day turn out to be very little more than dogma or the result of blind inertia. Should we overlook, for instance, the fact that Riabouchinsky's Koutchino Institute of Aerodynamics near Moscow had been in existence since 1904? Among many other pioneering activities, it studied the theory and aerodynamic practice of autorotating bodies and rotating cylinders, with and without tip discs. Riabouchinsky's own papers‡ and the official publications of the Institute are full of proofs of this statement.

On the basis of these proofs, I should like to make additional factual corrections to the history and philosophy of cylindrical rotors and autorotating bodies. The Koutchino Institute considered the three classical flows (Figure 95): (a) plane-parallel potential flow past a cylinder, with no circulation, (b) purely circulatory potential flow around the same cylinder, and (c) combined potential flow past the same cylinder (superimposition of the first two flows). Riabouchinsky and his colleagues carried out theoretical calculations, under N. E. Zhukovsky's supervision, of the aerodynamic characteristics of the combined flow, and then proceeded to the experimental study of the cylinder. Those who attended the tremendous reception given to Riabouchinsky in the Great Hall of the Northampton College of Advanced Technology, London (now The City University) on 9th January, 1961, may recall the public statement I made in the introductory speech. Here it is: '. . . If we now turn to the history of the so called Magnus Effect, which should, really, be called the Benjamin Robins Effect, we come up against yet another surprise: with all my love and respect for Professor Jacob Ackeret, who happens to be one of the closest friends of Riabouchinsky, I feel obliged, I feel in duty bound to say, that it was Dimitri Riabouchinsky,

† See, for instance, *Application of the Magnus Effect to the wind propulsion of ships.* L. Prandtl, NACA TM 367, 1926. *Der Magnus-Effect, die Grundlage der Flettner walze, z.der Deutsch. Ing.* A. Betz, Bd. 69, Nr. 3, January 3, 1925.

‡ See, for instance, *Etude sur l'hypersustentation et la diminution de la résistance à l'avancement, par Dmitrii Riabouchinsky.* Publications Scientifiques et Techniques du Ministere de l'Air, SDIT, Paris: 1957. Also *Thirty years of theoretical and experimental research in Fluid Mechanics.* D. P. Riabouchinsky, the Royal Aeronautical Society, London: 1935.

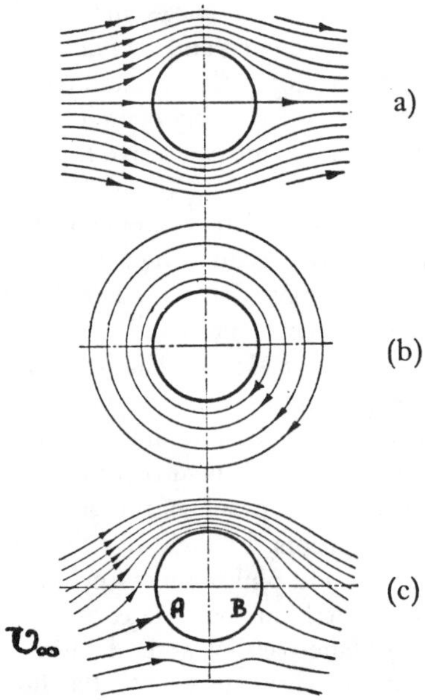

Fig. 95. The three classical flows around a circle

and not Ackeret, who studied the aerodynamic characteristics of rotating cylinders first, nearly fifteen years earlier . . .'

Professor Ackeret was, of course, informed by Riabouchinsky about this statement in advance. Moreover, being an old friend of Riabouchinsky, he was among the first to receive his (Riabouchinsky's) publications, which contained clear-cut passages like this one:†

'In Figure 96 are given the author's (i.e. Riabouchinsky's) experimental results for a single-winged rotor (1909) and Ackeret's experimental results (1924) for circular cylinders, with and without protecting discs . . .'

It may, of course, be argued that Riabouchinsky's curves represent autorotating bodies, while Ackeret's curves are for rotating cylinders. But, first, the fundamental theory is essentially the same in the both cases; second, Riabouchinsky studied also rotating cylinders, and Ackeret studied also autorotating bodies; third, Ackeret himself never dreamed that he was fully familiar with Koutchino experiments and theoretical investigations before he embarked on his own experiments.

What made Riabouchinsky undertake such research back in 1908–9,

† Ibid.

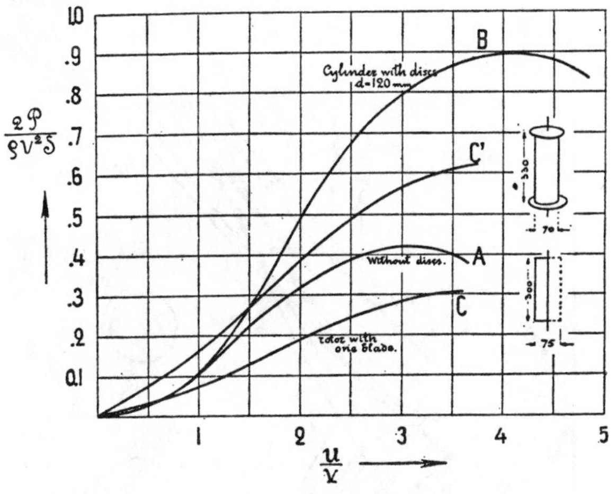

Fig. 96. Riabouchinsky's and Ackeret's experimental diagrams

or even earlier? The reasons were straight-forward. First, the Magnus Effect was in great fashion at the beginning of this century. Second, the Zhukovsky, or Kutta-Zhukovsky, theorem was still so fresh and so sensational that combined flows past cylinders (Figure 95c) occupied the minds of investigators. Third, Zhukovsky and Riabouchinsky were close friends and worked together, therefore it was only normal that the first should ask the second to carry out experiments on rotating cylinders, to check the Zhukovsky theorem experimentally. Fourth, the Russian department of Artillery requested the Koutchino Institute to study the behaviour of artillery shells in flight.

'I was very much impressed by the theories advanced by Frederick W. Lanchester', Riabouchinsky told me in 1963, 'and I must confess that I made a rather full use of his basic scheme of a bullet in flight (Figure 54, p. 98). I admired that man for the simplicity and clarity of his theories . . .'

May I be allowed to say once again, that an extensive programme of theoretical and experimental investigations of rotating cylinders, wing rotors and combined rotors was carried out by the author of this book in 1938–41. The main results for ordinary rotating cylinders emerged in the form shown in Figure 97. It should be noted that at certain (small) aspect ratios and certain (low) Reynolds numbers the 'Magnus Effect' changes its sign from plus to minus: a new 'effect' of basic aerodynamic interest. Further extensive experiments (in the Department of Aerospace Engineering of the University of Kansas, Lawrence, Kansas, USA, by

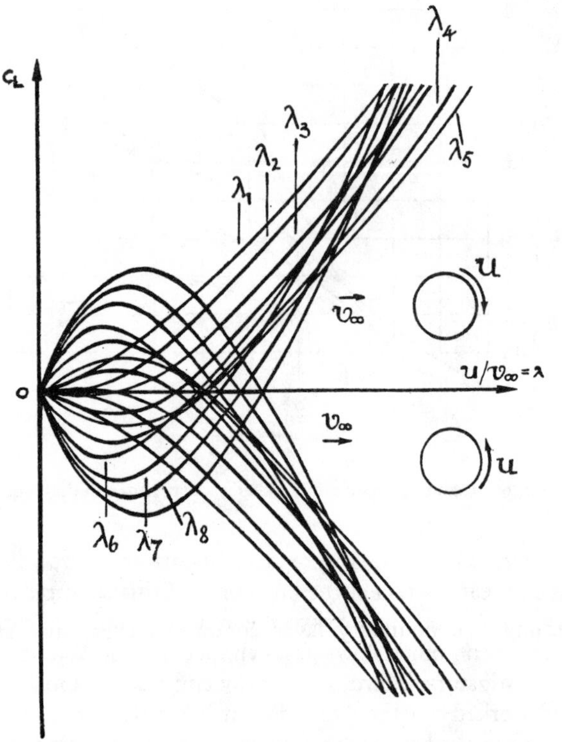

Fig. 97. Tokaty's aerodynamic characteristics of rotating cylinders

D. Lawrence in 1960, and in the Department of Aeronautics of the City University, London, in 1966–8) confirmed the new 'effect' beyond doubt.

Riabouchinsky, Mallock, Benard, von Karman

The extraordinary thing about Fluidmechanics is that whatever configuration or machine you invent, there is no escape from fluidmechanic resistance to motion. The component of the resultant force of this resistance in the direction perpendicular to the direction of motion is called Lift, or Lifting force, L; and its component in the direction opposite to the direction of motion is called Drag, D. Now, the latter consists, generally, of four parts: (1) form or pressure drag, (2) frictional drag, (3) induced drag, and (4) wave drag.

First, the form or pressure drag. D'Alembert and Euler showed that a sphere in a steady parallel flow of an ideal fluid experiences no resultant pressure drag; and Lord Rayleigh showed that the d'Alembert-Euler paradox applies also to a circular cylinder. The situation is, however, very different when they or any other body move in a real fluid.

The principal differences between the drag characteristics of one and the same body moving in ideal and real fluids (at relatively low speeds) are due to the absence and presence of viscosity, respectively. We shall see later on that viscosity is responsible for a thin layer of retarded flow, called the boundary layer, adjacent to the surface of the body. We have in this layer fluid shearing forces due to the skin friction on the body. At the same time, the boundary layer causes a flow pattern which disrupts the static pressure equilibrium predicted by Euler, d'Alembert, Stevinus, Pascal and Rayleigh, e.g. pressure forces cease to be normal to the surface, and, hence, a force called *form drag* emerges.

Obviously, the larger the surface covered by the boundary layer, the larger the resultant skin friction drag; and this makes streamlined bodies high skin friction drag bodies. Non-streamlined (or bluff) bodies, however, produce higher *form drag*. An ordinary flat plate is uniquely convenient to illustrate the two cases. When it is perpendicular to the flow (Plate 4), something like boundary layer conditions exist only at the tips, the surface of friction is small, therefore its frictional drag is very small. But the plate obstructs the oncoming flow with the whole of its frontal area, creates a large wake, and therefore the form drag is large.

Now, how do we know these facts? The answer is that almost every laboratory of aerodynamics has qualitative and quantitative methods and means for investigation. For example, Professor Riabouchinsky used to cover a sheet of iron with black powder sprinkled with lycopodium powder, and in this way visualized the flow patterns (Plates 11 and 12). The powder is blown off at points where speed or turbulence are maximum and the black paper re-appears. Lycopodium flow patterns obtained by this method popularized Riabouchinsky to such an extent that already in 1908–10 he began to receive letters and enquiries even from faraway countries. In 1962, the late Sir Frederick Handley Page told me that the first thing he knew about aerodynamics was Riabouchinsky's aerodynamic spectra (patterns).

'When I showed my flow patterns to N. E. Zhukovsky', Riabouchinsky recalled, 'he got excited and asked whether I could show also his bound vortex; I replied that even that could be done. . . .' 'Visualisation of flows brought me, however, to yet another curious phenomenon', he continued,† 'namely if at the moment when the wind begins to carry up

† There are numerous publications where this information can be found: *Le rôle de la*

the fine powder, one taps only once on the edge of the horizontal plate with a small hammer, the streamlines of the flow and other details are at once traced by the powder.' (As can be seen in Plates 11–12.)

These, and many other spectra, show, quite clearly, that an infinite mass of fluid is carried behind the bodies. Riabouchinsky used to call it a 'bound mass'. He showed me his old sketches on which he had tried to elaborate the kinematic structure of the 'bound mass'. I must say, however, that none of them looked to me like the so-called Karman Vortex Street: the only point of science on which I openly disagreed with my dear friend.

Look now at Plate 13, which shows a picture taken by a stationary camera of a circular cylinder moving to the left through water originally at rest. We observe a double row of alternating vortices following the cylinder. The vortices in the upper row are turning clockwise, while those in the lower row are turning counterclockwise. This system of vortices replaces the infinite mass of fluid assumed to follow the body in the theory of Kirchhoff, Rayleigh and Riabouchinsky. The surface of discontinuity can be considered as vortex sheets, and one finds, in general, that such vortex sheets are unstable, and they have the tendency to roll up so that the vorticity concentrates around certain points.

All the textbooks in Fluidmechanics call this *Karman's Vortex Street*, or Karman's Vortex Trail. But like Bernoulli's Equation or Magnus Effect they call it so inaccurately. Just read what Theodore von Karman himself wrote† on the subject:

'I do not claim to have discovered these vortices; they were known long before I was born. The earliest picture in which I have seen them is one in a church in Bologna, Italy, where St. Christopher is shown carrying the child Jesus across a flowing stream. Behind the saint's naked foot the painter has indicated alternating vortices. Alternating vortices behind obstacles were observed and photographed by an English scientist, Henry Reginald Arnulpth Mallock (1851–1933),‡ and then by a French Professor Henri Benard (1874–1939).§ Benard did a great deal of work on the problem before I did, but he chiefly observed the vortices in very viscous fluids or in colloidal solutions and considered them more from the point of view of experimental physics than aerodynamics. Neverthe-

---

méthode autorotationnelle et de celle des spectres aérodynamiques dans la mise en évidence du principe de l'aile à fente. Institut de France, Des Coptes: 1957. *Publications Scientifiques et Techniques du Ministère de l'Air.* N 167, 1970, and N 337, 1957, etc.

† *Aerodynamics.* Theodore von Karman, Cornell University Press, Ithaca: 1957.
‡ *On the resistance of air*, series A, 79, 1907, pp. 262–72. A. Mallock, London.
§ *Formation de centres de giration à l'arrière d'un obstacle en mouvement*, Comptes rendus de l'Académie des Sciences, Paris, pp. 839–42, 970–2. H. Benard, Paris: 1908.

less, he was somewhat jealous because the vortex system was connected with my name, and several times, for example, at the International Congress for Applied Mechanics held in Zurich (1926) and in Stockholm (1930) he claimed priority for earlier observation of the phenomenon. In reply I once said, "I agree that what in Berlin and London is called 'Karman Street' shall in Paris be called 'Avenue de Henri Benard'." After this wisecrack we made peace and became quite good friends.'

What, then, did von Karman contribute to the vortex street theory? He was the first to show that the symmetric arrangement of vortices (Figure 98) is essentially unstable; that only the assymetric arrangement,

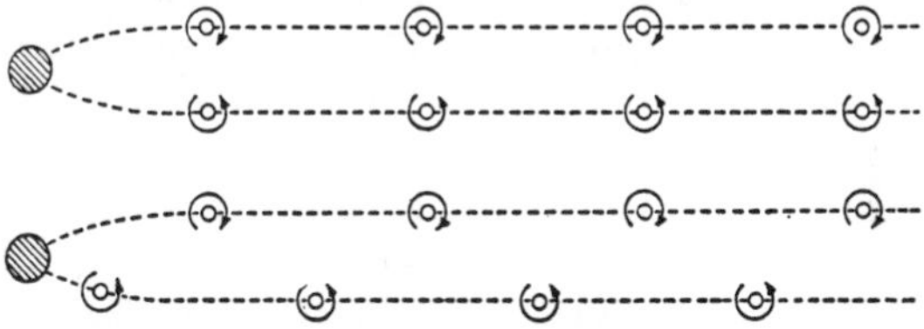

Fig. 98. Symmetric and asymmetric rows of vortices

(lower row of figure) is stable, but only for certain ratios of the distance between the rows and the distance between two consecutive vortices of each row, and that the momentum carried by the vortex system is connected with the drag of the body – he showed, in fact, how such a vortex system can represent the mechanism of formation of the wake drag:

$$D = \rho l v^2 \left[0{\cdot}314 \left(\frac{u}{v}\right)^2 - 0{\cdot}794 \left(\frac{u}{v}\right)\right],$$

where $u = -\gamma/l\sqrt{8}$, γ the circulation of a single vortex, l the distance between the neighbouring vortices of a row.

William Froude (1810-79) and others

The laws of fluidmechanic resistance, as we know them today, were established and brought into being stage by stage as the logical consequence of the efforts by Leonard Euler, Daniel Bernoulli, Pascal, Mariotte, Newton, Borda, Bossut, Du Buat, d'Alembert, Riabouchinsky, Robins, Zhukovsky, and others.

But when shipbuilding entered the era of science and enlightened technology, and the experimental techniques began improving rapidly, it became clearer and clearer that the role played by friction in the formation of drag could no longer be neglected. Saint-Venant, Kleitz, Helmholtz, Meyer,† Coulon,‡ Couplet and their contemporaries, each in his fashion, tried to show, both theoretically and experimentally, that viscosity was the all-important factor of fluidmechanic resistance. Towards the end of the last century, it became an accepted philosophy that the laws governing fluid friction were different from those of simple (Newtonian) viscous resistance, because of the fact that, except at extremely low speeds, the motion of the fluid becomes unsteady, eddies are formed, and the energy absorbed in fluid friction is spent chiefly on the formation of vortex eddies.

In 1872, an English engineer and naval architect called William Froude (1810–79), carried out a series of experiments at Torquay for the Admiralty. Plates and boards of differently prepared surfaces were towed endwise through the still water of a large basin. The carriage was fitted with a dynamometer and automatically recorded the velocity and drag of the boards.§ His main conclusions were that:

(1) The fluidmechanic resistance varies greatly with the condition of the surface: (2) it is proportional to v^n, where n depends on the state of the surface (n decreases, up to a certain limit, with an increase of length, but is independent of the velocity): (3) the total fluidmechanic resistance increases with the length, though the resistance per square foot decreases as the length increases.

Owing to viscous drag those parts of the surface near the prow communicate motion to the water, so that the relative motion is smaller over the rear part of the surface and the drag per unit area is consequently less. The velocity of the accompanying flow increases until at some point on the surface a balance is reached between the amount of energy given to the accompanying stream per second, and the energy dissipated by eddy formation in the surrounding fluid and in producing motion of a greater volume of this water against viscous resistance. After this point is reached, the velocity of the flow and the resistance per square foot of surface remain approximately constant, he concluded.

Froude's investigations were, in fact, wider than the towing of plates. For example, he participated in the design, construction and calibration of the famous Heenan and Froude Hydraulic Dynamometer, which made his and others' experiments possible. He also carried out theoretical investigations, one of the better known results of which was the so-called Froude Number, $F_r = v^2/lg$, l being a characteristic length of the body,

† *Über die Reibung der Flussigkeiten*. O. E. Meyer, 'Poggend. Ann', B.113, 1861, S.400: Berlin.

‡ *Mémoires de l'Institut de France*, t.III, 1813, p. 282, Paris.

§ See *Inst. Naval Architects*. March, 1898. Also *Hydraulics and its applications*. A. H. Gibson, Constable & Co Ltd, London: 1908.

and g acceleration of gravity. This dimensionless number has generally little aerodynamic interest, but it is of considerable importance in ship design, where gravitational (wave) forces are the primary determinant of the total force.

We shall meet Froude also in the section dealing with airscrews. Figuratively speaking, what he did in water, a German engineer called Zahm did in the air.† Zahm was, perhaps, the first man in the history of aerodynamics to undertake a programme of study of the nature and quantitative laws of air friction on wind-tunnel walls. The results obtained by him were similar to those by Froude, but the absolute values were, of course, proportional to the mass densities of water and air, respectively. If the ratio of the kinematic viscosities of water and air is 13 : 1, he said, then a velocity of 10 f.p.s. with the 4-feet board in water must correspond to a velocity of 32·5 f.p.s. with the 166-feet board in air. Thus, taking the resistance in air proportional to $v^{1\cdot 85}$, the resistance per square foot of the 16-feet board at this speed had to be $0\cdot 000457 \times (3\cdot 25)^{1\cdot 85} \leqslant 0\cdot 00402$ pounds, and the $R \propto v^2$ law gives 0·00325 for the 4-feet board in water, the ratio of the two being 855, a value of about 4 per cent greater than the relative density of water and air at 60°F.

The rapidly growing practical needs required, however, still greater accuracy, and this was soon provided by Ludwig Prandtl, the creator of the boundary layer theory.‡ What Froude and others tried to solve experimentally, he resolved theoretically and experimentally. He postulated that the fluid particles immediately adjacent to the solid surface (of Froude's boards or of other bodies) remain in the non-slip state, i.e. are motionless (in relation to the surface). But as the distance from the surface increases, the particles move faster and faster. At a certain distance δ, called 'boundary layer thickness', they move with the same velocity as the flow outside δ. The character of change v within the boundary layer is shown in Figure 99.

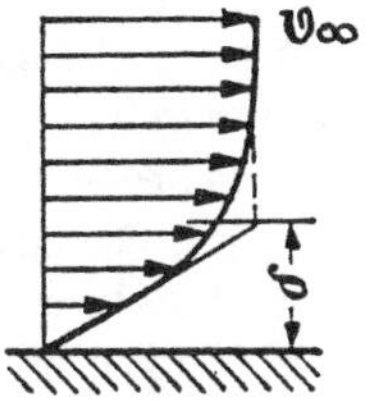

Fig. 99. The boundary layer velocity profile

† Phil. Mag., vol. 8. 1904.

‡ *Ueber Flussigkeits bewegung bei sehr kleiner Reibung*. L. Prandtl, Berhandl. d. III, Internat. Mathem. Kongr., Heildelberg, 1904.

I think I should add here that Prandtl was not actually the first to attempt to solve the integro-differential equations of viscous flows, the Navier-Stokes equations. Back in 1851, Stokes succeeded in solving them (for the motion of a sphere) and obtained his famous formula for the drag experienced by a sphere of radius r_0, $D = 6\pi\mu v_\infty r_0$. Nor was he the first to discover the existence of the velocity profile. For example, Hagen wrote† that when one watches the flow of coloured water in a glass pipe, it becomes obvious that, at low pressures and temperatures, the fluid flows in extremely thin concentric-coaxial cylinders, one inside another, parallel to each other. We should mention also Poiseuille who proved theoretically that these (Hagen's) cylinders move with different speeds, their velocity profile across the pipe being a parabola with the maximum velocity along the axis of the tube.

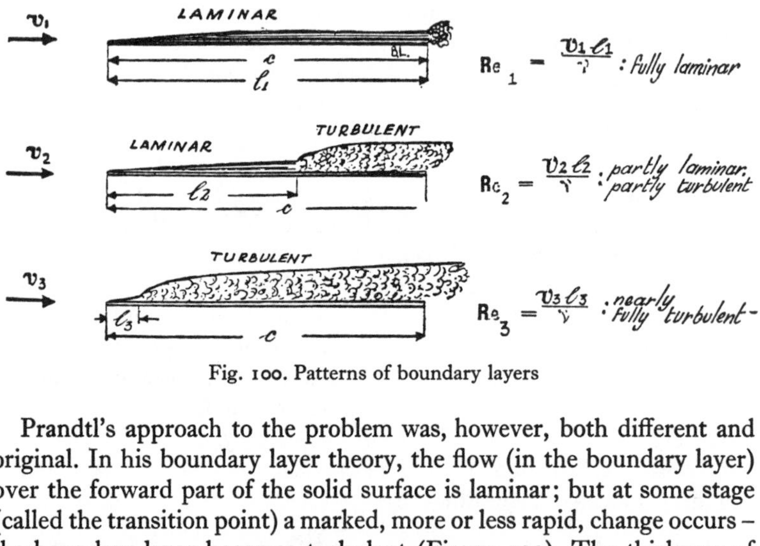

Fig. 100. Patterns of boundary layers

Prandtl's approach to the problem was, however, both different and original. In his boundary layer theory, the flow (in the boundary layer) over the forward part of the solid surface is laminar; but at some stage (called the transition point) a marked, more or less rapid, change occurs – the boundary layer becomes turbulent (Figure 100). The thickness of the turbulent boundary layer grows more rapidly than that of the laminar boundary layer, and because the eddies or turbulent fluctuations provide an effective mixing mechanism, the velocity distribution in it is more uniform than in the laminar boundary layer.

Experimental and theoretical studies of the boundary layers, apart from Prandtl's, have been going on for very many years in very many countries. Probably one of the fullest and best publications on the subject was L. G. Loytsiansky's 'Aerodinamika Pogranichnogo Sloya' (Boun-

† Abhandl. der Kgl. Academie der Wissenschaften zu Berlin, 1869, 2B, S. 1 and 2.

dary Layer Aerodynamics), Moscow, 1941.† But all researchers in the field would agree that, however beautiful the mathematical results they obtained, some aspects of the problem require further study. Nevertheless, the achievements are outstanding and some of the working formulae emerged in forms simple enough.

The solution of the differential equations of motion of viscous fluids proved to be exceptionally difficult. For this reason, Blasius considered an infinitely thin plate of length L in a uniform incompressible flow of velocity v_0, density ρ and viscosity μ, and attempted to obtain an approximate solution not for the finite length L but for the infinitely long distance ox, of which L was a part. He then used the solution for the distances $0 \leqslant x \leqslant L$. The stream velocity v_0 and therefore also the pressure p were assumed to be constant along the boundary; at the leading edge of the plate the thickness of the layer was assumed to be zero. For these conditions, he obtained for the laminar boundary layer (flat plate):

thickness .... $\boxed{\delta_L = \dfrac{4{\cdot}64x}{\sqrt{Rex}}}$,

shearing stress .... $\boxed{\tau_L = 0{\cdot}332\sqrt{\dfrac{\mu\rho v_0^3}{x}}}$,

friction drag coefficient .... $\boxed{C_{fL} = \dfrac{0{\cdot}664}{\sqrt{Rex}}}$,

where *Rex* is the local Reynolds number, x being the distance from the leading edge.

Thus, the determination of the (viscous) frictional part of the fluid-mechanic resistance became a reality. Indeed, knowing C_{fL}, the whole frictional drag can be computed by the now classical formula

$$\boxed{D_{fL} = 2\int_0^L \tau_L dx = 2 \times \frac{C_{fL}\rho v_0^3}{2} = 1{\cdot}328\sqrt{\mu\rho L v_0^3}}$$

It should, however, be remembered that Blasius' approximations had introduced certain simplifications; for example, it is impossible to create an infinitely thin flat plate. But, still, the above results meet the essential needs of engineering practice.

† The essential results of this valuable work reappeared in Loytsiansky's book *Mekhanika Zhidkosti i Gaza* (Mechanics of Liquids and Gases). Moscow: 1959 (in Russian).

Turbulent boundary layer and flow separation

As to the turbulent boundary layer, it is a very much more complicated phenomenon demanding special study and experimental treatment. A good deal of basic information on the subject was provided by Osborne Reynolds,† but more information was needed. Theodore von Karman suggested a unified boundary layer theory to embrace the boundary layer as a whole, instead of trying to solve the partial differential equation,‡ and Karl Polhausen showed the advantages of such a theory.§

These and other efforts resulted in the following working formulae for the turbulent boundary layer:

$$\boxed{\delta_T = \frac{0{\cdot}371\, x}{\sqrt[5]{Rex}}}$$

$$\boxed{\tau_T = 0{\cdot}0225 \rho v_0^2 \left(\frac{\nu}{v_0 \delta_T}\right)^{\frac{1}{4}}}$$

$$\boxed{C_{fT} = \frac{0{\cdot}072}{Rex^{0{\cdot}2}}}$$

There are also formulae for the mixed boundary layer, and various modifications of the formulae. Almost all of them involve the Reynolds number, and this underlines the historic importance of Osborne Reynolds' investigations. But the problem of turbulence goes far beyond the boundary layer. Water and air flows past natural obstacles, boats, ships, swimmers, houses, land vehicles, bridges, etc., as well as flows in pipes, tubes, tunnels, chimneys, etc., are not free from turbulence. In fact, one could go as far as to suggest that there are no real fluid flows free from turbulence.

Let us, therefore, consider the behaviour of fluid particles more generally. They arrive at the leading edge with the kinetic energy

† Phil. Trans. A, 174, 1883 and 1895.

‡ *Über laminare und turbulente Reibung*, ZAMM, pp. 233–52: 1921.

§ *Zur näherungsweisen Integration der Differentialgleichungen der laminaren Grenzschicht*, ZAMM, pp. 252–68. 1921.

$mv^2/2$, or $\rho v^2/2$. But as soon as they enter the boundary layer, they find themselves against the viscous friction resistance. Its overcoming absorbs a part of the kinetic energy. And so the $mv^2/2$ or $\rho v^2/2$ is becoming less and less, the motion becomes slower and slower, each of the particles becomes more and more 'tired'. Then, at some point, the $mv^2/2$ or $\rho v^2/2$ is so small, that the particle can no longer continue its orderly motion, it first stops or nearly stops, then separates from the surface, from the boundary layer.

And that is not all. If we take, for example, a semicylinder (Figure 101), we see that the cross-sectional area of the flow in the region of A is

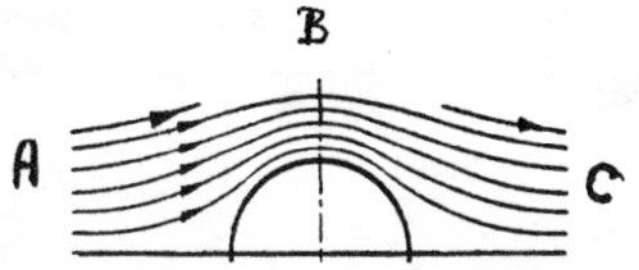

Fig. 101. Potential flow past a semi-cylinder

large, therefore, according to Leonardo's law vA = const, the velocity v is relatively small; which means that, by Lagrange's integral, pressure p is relatively high. When particles proceed towards the region B, the cross-sectional area A decreases, therefore v increases and p decreases. Now watch carefully the further movement: in the region of C the cross-sectional area is again large, therefore v is relatively small and p relatively high, i.e. $p_B < p_C$, which means that from B to C particles move against higher pressures. When there is no viscosity, the kinetic energy gained from A to B is exactly enough to overcome the increasing pressure from B to C. But in real flows the particles do not gain as much of kinetic energy (from A to B) as in an ideal flow (because of viscous friction), and their way from B to C is not free from viscous resistance.

Thus, at some point, the particles stop and separate. Behind the point of separation the mass of fluid moves backwards (Plate 14).

Unfortunately, we cannot go into the details of these extremely interesting phenomena. But we cannot fail to say that almost as soon as the facts of transition and separation were established, man started 'fighting' them. Betz, Ackeret, Flettner, Alexandre Favre, D. P. Riabouchinsky and others attempted to develop methods of delaying or preventing flow separation.

Methods of delaying flow separation

Professor Riabouchinsky once said, and this can be seen in his publications,† that when he read Prandtl's boundary layer theory, his immediate reaction was 'to invent a method of prevention of formation of such a layer', and thereby 'to exclude all its nasty consequences for fluid flows'. He began working on his device – a moving surface – in 1912. In a paper published in 1914, the idea was developed by him to the extent revealed in Figure 102. This gives the basic idea (Riabouchinsky

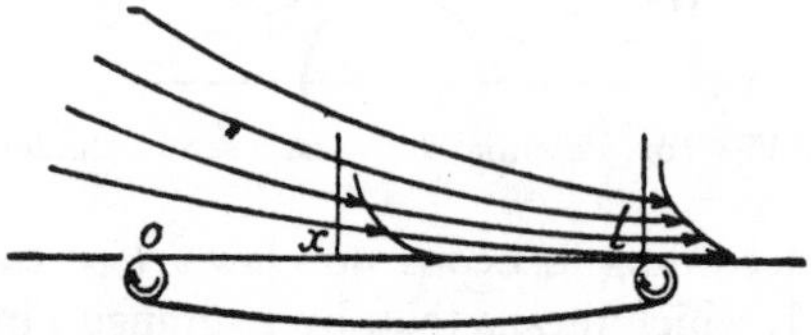

Fig. 102. Riabouchinsky's moving ground scheme

also published its mathematical-aerodynamic theory), while Plate 16 shows the actually constructed and used running surface, which he called 'an endless belt'. Together these demonstrate the arrangements used to measure the frictional resistance by the momentum of the air swept off by the belt. He not only succeeded in preventing the formation of the boundary layer, but changed the sign of the resistance, and made it an accelerating force.

As has already been said, Anton Flettner intended to use this device instead of a single rotor, hoping that it would give him a sharp increase in the aerodynamic circulation and, consequently, of the rotor force itself.

The Göttingen Centre of Experimental Aerodynamics tried to solve the problem of prevention of early flow separation by means of removing the low-energy parts of the boundary layer, or of adding kinetic energy to the boundary layer. This could, and can, be done by removing low-energy air through suction slots or a porous surface. Another common method is to blow high-energy air through backward-directed slots (Figure 103). The air handled through either the suction or blowing slots

† *Etude sur l'hypersustentation et la diminution de la résistance a l'avancement.* D. Riabouchinsky, SDIT, Paris: 1957.

may be carried through the interior of the wing. By far the most extensive investigations along these lines were carried out by Schrenk.† Using a single sucking slot, he succeeded in preventing flow separation from a thick wing and thus in increasing its maximum lift coefficient up to

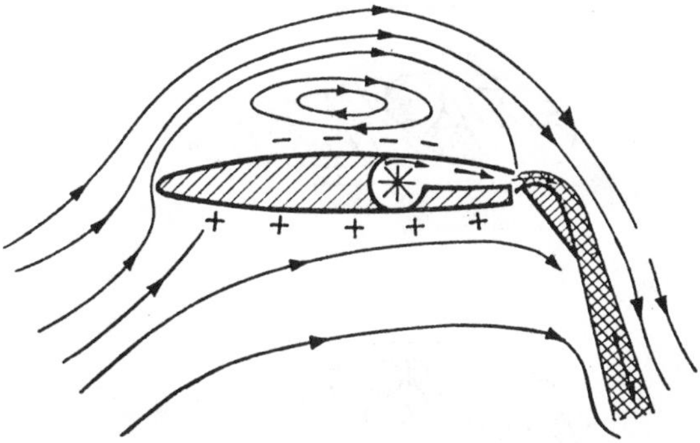

Fig. 103. A method of blowing high-energy air through backward slots

about 5, nearly three-times. Similar experiments were also going on in NACA (USA), CAGI (USSR) and other research establishments. The CAGI even built an aircraft (approximately 1935–6), with a full-scale boundary layer sucking in flight.

In about 1925–7, B. N. Your'yev of the USSR proposed a wing with wholly moving surfaces. Ten years later, on his insistence, I undertook the theoretical and experimental study of a similar wing: 1 metre span and 22 cm chord (Figure 104). It was studied in the T-1 wind-tunnel

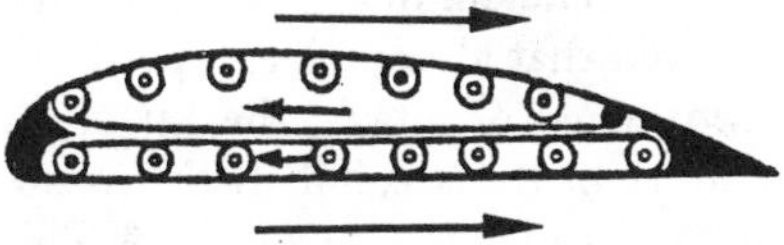

Fig. 104. B. N. Your'yev's wing with moving surface

of the Zhukovsky Academy of Aeronautics, Moscow, but proved to be a technically unsatisfactory device. We then built simpler wings with fully or partly moving surfaces, some of which worked quite well and delayed the flow separation up to about 25° of angle of attack (Figure 105). In

† Experiments with a wing from which the boundary layer is removed by suction, NACA TM No. 634, 1931, TM No. 773, 1935, and TM No. 534. Oskar Schrenk: 1929.

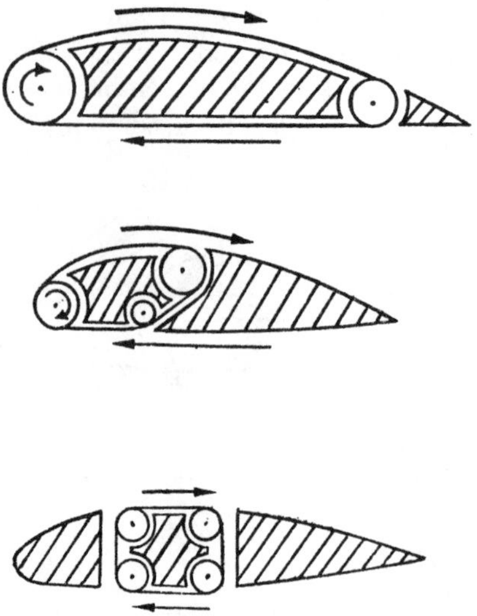

Fig. 105. Tokaty's wings with moving surfaces

Plate 17, we have Alexandre Favre's (Marseilles) wing, which delayed the separation up to 105° of angle of attack.†

Finally, a few words about the so-called high-lift devices. In order to understand this work, we must now introduce a new name: Sergei Alekseyevich Tchaplygin, or Chaplygin (1869–1942), a close friend and collaborator of Zhukovsky, one of the more outstanding mathematicians and aerodynamicists of Russia and the USSR. His contributions to theoretical Fluidmechanics were of a fundamental nature and remain to this day among the upper echelons of achievement. In his mathematically rigorous style, he showed that at any given speed of flight v_∞, altitude (ρ = const), and zero angle of attack, the lift of an aerofoil depends entirely on the curvature of the aerofoil. If the mean camber-line is an arc (Figure 106), the lift formula is $L = 2\pi\rho h v_\infty^2$. This theorem explains

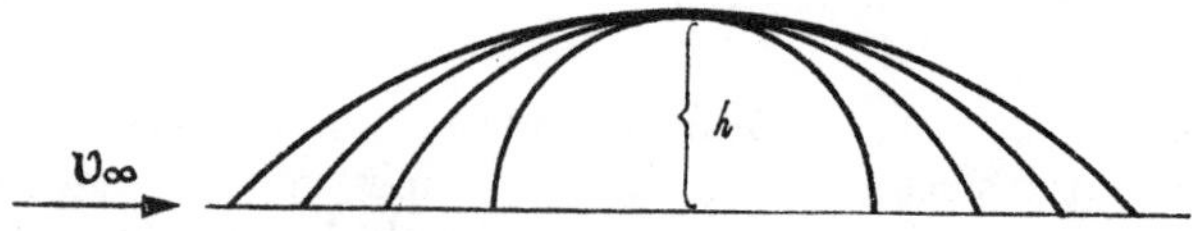

Fig. 106. Chaplygin's theorem

the usefulness of the various high-lift devices incorporated in a modern aircraft, some of which are shown in Figure 107.

† *Krylo s dvizhuscheisya poverkhnost'yu.* G. A. Tokaev, 'Samolyet', Moscow: 1940.

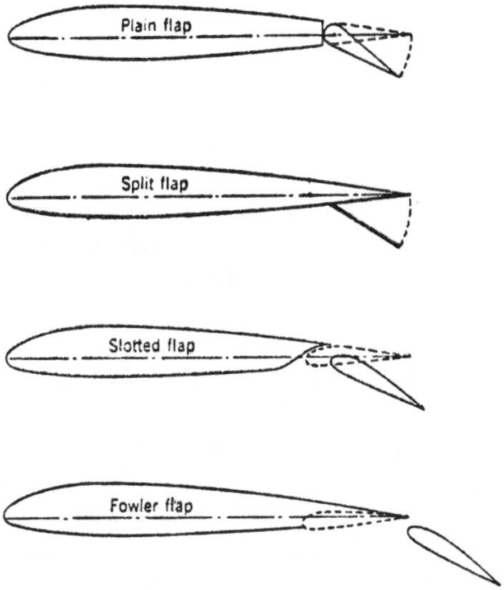

Fig. 107. Typical high-lift devices

The *plain flap* is formed by hinging the rearmost part of the aerofoil about a point within the contour. Downward deflections of the flap effectively change the camber of the aerofoil, therefore its lifting capacity improves (we could say that the flow separation is delayed).

The *split flap* is formed by deflecting the aft portion of the lower surface about a hinge point on the surface at the forward edge of the deflected portion, and produces a large increase of camber, thereby improving the maximum lifting capacity of the wing.

The *slotted flap* provides one or more slots between the main portion of the wing section and the deflected flap; and they derive their effectiveness from increasing the camber. The slot(s) duct high-energy air from the lower surface to the upper surface and direct this air in such a manner as to delay separation of the flow over the flap by providing boundary layer control. There are numerous types of slotted flaps.

The *leading edge slats* are small aerofoils mounted ahead of the leading edge of the wing in such an attitude as to assist in turning the air around the leading edge at high angles of attack and thus delay leading-edge stalling. They may be fixed or retractable.

Complete information on these and other high-lift devices can be found in almost every textbook in Aerodynamics and in special monographs.†

† See for example, 'Theory of Wing Sections'. Ira H. Abbott and Albert E. von Doenhoff, McGraw-Hill Book Co, Inc., New York: 1949.

Airscrews

We thus see that every real body moving in any real fluid experiences a definite fluidmechanic resistance of one kind or another. How can this drag, or resistance, be overcome? How can the motion be created and maintained?

I have already referred to Hero's jet machine, Leonardo da Vinci's version of Archimedes' waterlifting screw, Lomonossov's helicopter with counter-rotating blades, Rykatchev's spring-driven propeller, etc. All these were the grandfathers of modern propulsion systems, which can today be grouped into three categories: Airscrews, Jet Engines, and Rocket Motors. They all employ fluidmechanic principles.

An airscrew, or an air propeller, is a device for propelling a craft through the atmosphere. It has blades which work roughly speaking, as aircraft wings, and produce the force needed to overcome both the air resistance and the inertia of the craft. The classical airscrew-propeller consists of a hub and two or more blades. The hub is so designed that it can be mounted on and rotated by an aeroengine, thereby imparting to the blade's rotational motion in the plane nearly perpendicular to the direction of motion of the craft (Figure 108).

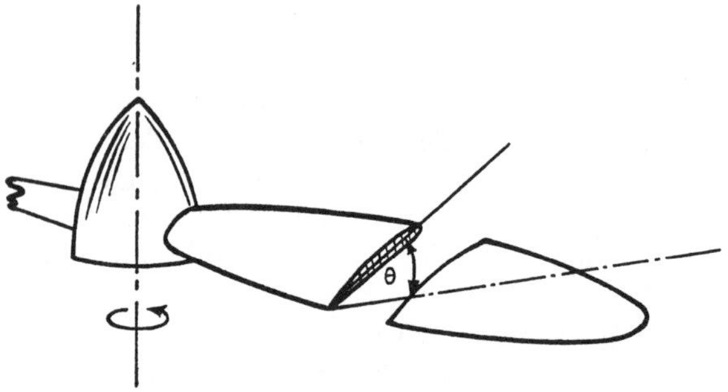

Fig. 108. Basic airscrew scheme

The theory of airscrew propulsion was initiated by W. J. M. Rankine in England and others.† Zhukovsky's theory was based on the so-called

† (Transactions of the Institute of Naval Architects, Vol. 6, 1865), N. E. Zhukovsky in Russia (Coll. Works, Vol. VI), O. Flamm in Germany (Die Schiffschraube und ihre Wirkung auf das Wasser, 1910), R. E. Froude in England (Transactions of the Institute of Naval Architects, Vol. 30, 1889).

'Vortex Trails', and became known as the Vortex Theory. Rankine's theory was based on the fluidmechanic momentum principle; it dominated the line of thought for a long time, and it therefore deserves a fuller description.

Although an airscrew has a limited number of blades, it is assumed, for simplicity, that the number of blades is infinite so that the thrust is axially symmetric and evenly distributed over the whole disc swept out by the blades. The screw-propeller can then be regarded as a device which provides a pressure jump, from p_1 to p_2, say, across the disc from front to rear. The pressure ahead of the propeller is assumed to be ambient, p_∞, say, and it must return to this value somewhere behind the disc in the ultimate slipstream (Figure 109). V. P. Vetchinkin (1888–

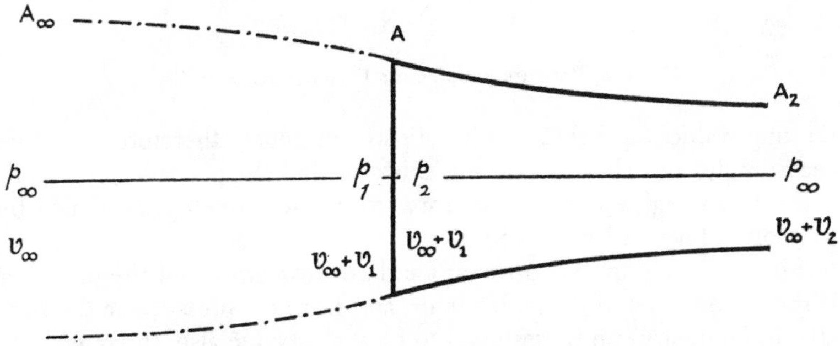

Fig. 109. Froude's idealised slip stream used for the development of the momentum theory of airscrew thrust

1950), one of the closest collaborators of Zhukovsky and, undoubtedly, one of the more outstanding aerodynamicists of the first half of the twentieth century (see, for example, *Collected Works* by V. P. Vetchinkin, the Academy of Sciences of the USSR, Moscow, 1956), once remarked that the greatest single shortcoming of the theory under discussion was 'the lack of reliable knowledge of what exactly takes place when air particles cross the disc'. Indeed, countless attempts to find out the answer showed that the kinematic and dynamic processes across the plane of rotation are so complicated that no precise mathematical treatment is possible.

Mainly for this reason, man had to introduce the 'ideal propeller' concept: an airstream of circular cross-section containing the propeller disc, within which the oncoming air particles accelerate. This concept is often referred to as Froude's concept, or Froude's Actuator Disc, with a jump of pressure as shown in Figure 110. It will be seen that there is positive pressure gradient upstream everywhere, which means that the fluid's speed increases steadily from far in front of the propeller to a

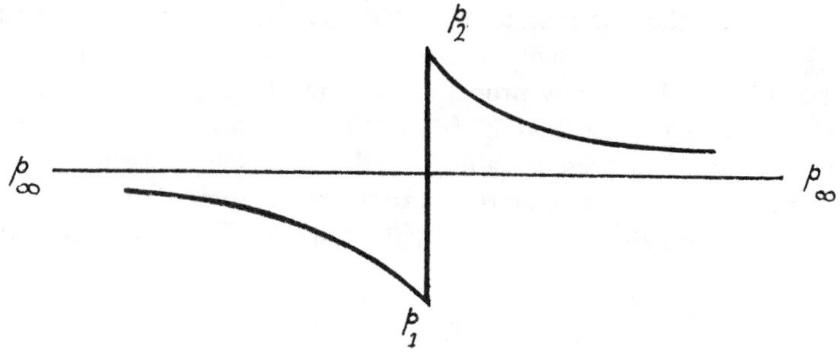

Fig. 110. Pressure jump across Froude's actuator disc

limiting value far behind. The slipstream must, therefore, contract accordingly, and although it only exists behind the propeller, it is convenient to imagine it extending forward of the propeller, as shown by the dotted lines in Figure 109.

Since the pressure far upstream and far downstream of the propeller is the same as that of the ambient air, and since the pressure on the slipstream boundary can be assumed to be at this value also, the only force acting on the air in the whole slipstream is that from the thrust on the disc, and this force is $(p_2 - p_1)A$. Now the mass of air passing in unit time in the slipstream can be conveniently evaluated at the propeller itself and is, therefore, $\rho A(v_\infty + v_1)$. The total increase in velocity of this mass of air is v_2, so that the rate of change of momentum of the slipstream is $\rho A(v_\infty + v_1)v_2$, therefore $p_2 - p_1 = \rho(v_\infty + v_1)v_2$. On the other hand, the application of the Bernoulli-Lagrange law to the flow in front of and behind the disc gives $p_2 - p_1 = \rho(v_\infty + v_2)v_2$. Comparing these expressions, we find that $v_2 = 2v_1$, and this is known as the Froude-Finsterwalder theorem. Thus, momentum thrust emerged in the form $T = (p_2 - p_1)A = 2\rho A(v_\infty + v_1)v_1$.

How efficient is the ideal propeller, the actuator disc, producing this thrust? The answer can be constructed as follows. The useful power is the thrust multiplied by the velocity v with which the propeller moves forward, $E_o = Tv_\infty$ and the power expended is the thrust times the airflow velocity through the disc, or $E_i = T(v_\infty + v_1)$. Comparing now the two powers, we obtain the so-called ideal propeller efficiency: $\eta = Tv_\infty/T(v_\infty + v_1) = v_\infty/(v_\infty + v_1)$.

It should, however, be pointed out that nowhere in this theory was any attention paid to such important facts as vortex formation, boundary layer friction, compressibility effects, etc. – air resistance generally – caused by the moving blades. For this reason, the ideal propeller efficiency is significantly higher than the efficiency of a real propeller. The other shortcoming of the theory is that it does not tell us anything about how the blades, the creators of the thrust, should be designed aerodynamically.

D. P. Riabouchinsky was among the first to stress these shortcomings, and to suggest remedies. The results of his theoretical and experimental investigations on the fluidmechanic characteristics of water- and airscrews were published in the 'Koutchino Bulletin' in January, 1909. At about the same time, namely in 1908, R. E. Froude gave an account of the results of his model experiments with steamship screw propellers. 'The theoretical formulae obtained by me and by Froude are in disagreement', wrote Riabouchinsky,† 'and so are also our experimentally determined coefficients; but there are also stronger objections against the application of Froude's formula in all cases. Namely, in studying the repartition of velocities in the slipstream behind a stationary airscrew of pitch h, the author (i.e. Riabouchinsky) never observed axial velocities of the rate $2nh$, n being the number of revolutions. The velocities actually observed by him tended towards nh and never exceeding this limit.'

Another approach to propeller theory, called the 'blade element theory', treated the blades as rotating wings. By knowing, or assuming, the aerodynamic characteristics of the aerofoil section of the blade element and taking into account the fact that the relative airspeed at the element depended on its radial distance and the forward speed of the propeller, one could compute its lift and drag. By integrating over all the elements of the blade and over all the blades, the thrust and torque of the propeller could be estimated.

It was realized by Drzewiecki that it would be wrong to use aerofoil characteristics determined from two-dimensional wind tunnel tests, but that some account should be taken of the fact that the aspect ratio (the span to chord ratio) of the blades is finite. The question arises: what aspect ratio should be used, because the blades operate under conditions different from a wing, so that tests made on wings of the same aspect as the blades may not provide valid data? Drzewiecki proposed a series of

† Those interested in the subject are referred to the following original sources: Bulletin de l'Institut Aerodynamique de Koutchino, December, 1909. Bulletin de l'Institut Aerodynamique de Koutchino, January, 1909. Etude sur l'hypersustentation et la diminution de la resistance a l'avancement, par Dimitri Riabouchinsky, Paris: 1940.

tests on special propellers to provide a mass of data for design purposes. One of the consequences of the blade element theory is that as the drag approaches zero, the efficiency of the propeller approaches unity, whereas the momentum theory, previously referred to, predicts an efficiency less than unity. This apparent discrepancy is resolved by the fact that the blade element and momentum theories, far from being in opposition to one another, should be used in conjunction, for each provides the information the other lacks.

The reconciliation of the blade element and momentum theories was pursued in many countries as a logical extension of wing theory which had been developed considerably by about 1920. Wing theory discussed in an earlier section, is founded upon the idea that lift is due to circulation around the wing and that, since the circulation, or the 'bound vortex', cannot end abruptly at the wing tips, free vortices spring from the wing and trail backwards behind it. The lift of a particular element of the wing is related simply to the local circulation, and the change of circulation from element to element determines the strength of these free vortices. In turn, the free vortices give rise to an induced velocity at the wing itself which contributes to the local flow direction and, thus, through a knowledge of the lift characteristics of the section, to an evaluation of the local lift. This description in words is expressed mathematically by two equations connecting the local lift and induced velocity which, when combined, result in the well-known 'integral equation' for the wing lift.*

Precisely the same approach was made in the case of the propeller by H. Glauert (1894–1934),† E. Pistolesi,‡ and S. Kawada.§ In the propeller case, however, the free vortices are not straight, as in the case of the wing, but lie, roughly, on the spiral paths traced out by the blade elements from which they sprang, giving rise to vortex sheets (Plate 18). If the propeller had an infinite number of blades, the vortex lines would coalesce to form vortex cylinders. There would then be radial symmetry and one of the interesting consequences of this is that the influence of the vortices springing from a blade element is confined to the annulus swept out by that element only, in sharp contrast to wing theory in which the corresponding vortices affect every other part of the wing and the total effect, at any given wing section, can only be obtained by an integration along the entire wing. The so-called 'independence of blade elements', however, greatly simplifies propeller theory by enabling the simpler momentum equation to replace the integral, referred to above, which occurs in the wing theory. Since the induced velocity

† British A.R.C. R & M 786 and 869, 1922.

‡ 'Vorträge aus dem Gebiete der Hydro- und Aerodynamik, Innsbruck: 1922.

§ Tokyo Imperial University, Aero. Res. Inst., No. 14, 1926.

* Pages 184 to 187 were written mainly by Dr. A.R.S. Bramwell, Senior Lecturer, Department of Aeronautics, The City University, London.

created by the thrust on an annulus occurs over the annulus only, the relationship between the velocity and the thrust can be obtained by considering the rate of change of momentum of the air through the annulus. This relationship has already been calculated for the complete disc. For an annulus of radius r and width dr, the corresponding area is $2\pi r dr$ and the Froude-Finsterwalder equation is usually written in the differential form

$$dT = 4\pi r(V + v_1)v_1 dr.$$

A propeller blade may be considered as a strongly twisted wing; this is necessary in order to ensure that each blade section operates at a favourable angle of attack α (Figure 111). Each individual blade section performs two simultaneous motions: it moves forward with the speed V of the aircraft, and rotates about its axis at the speed $u = \omega r = 2\pi rn$, with a resultant speed W, say, n being the number of revolutions per second.

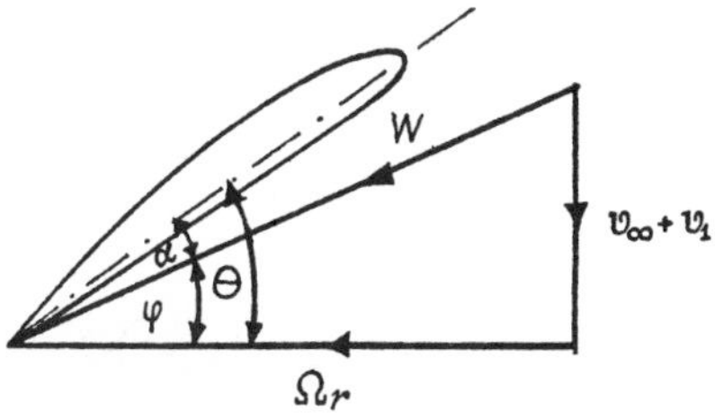

Fig. 111. The propeller blade element

Treating the blade section as the section of a wing and expressing the local incidence φ in terms of these axial and tangential velocities, the thrust on the annulus for b blades is

$$dT = \tfrac{1}{2}b\rho cW^2(C_L \cos \varphi - C_D \sin \varphi)dr$$

The 'inflow' angle φ clearly depends on v_1, so that the unknown v_1 can be eliminated between the two equations for dT which can then be calculated. In these forms the two equations combine mathematically the momentum and blade element principles mentioned earlier and which, formerly, could not be reconciled. Similar relations exist for the torque.

When the thrust T and torque Q of the whole propeller have been calculated the efficiency is found from

$$\eta = \frac{TV}{\omega Q}$$

The so-called 'advance ratio', $J = \frac{V}{nD}$, where n is the revolutions per second and D is the diameter of the propeller, is a scale factor, a constant value of which ensures that a series of propellers operates under geometrically similar conditions. The advance ratio is a measure of the distance travelled by the propeller during one revolution and of the helix angle of the vortex filaments of the propeller wake. If we calculate the thrust, torque and efficiency for a number of values of J and plot them against it, for a given blade setting angle, we have Figure 112.

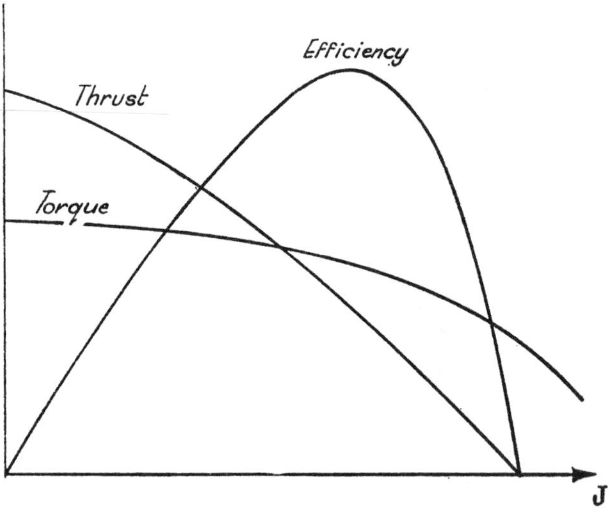

Fig. 112. Typical propeller characteristics

These curves depend on the blade setting angle and a series of curves is usually produced with, say, blade root angle as a parameter. Usually the rotational speed of the propeller is kept constant in flight so that J becomes a measure of the flight speed. It is obvious from the curves that the efficiency changes very considerably with flight speed if the blade angle is fixed and that it would therefore be advantageous to be able to change the blade angle so that maximum efficiency is achieved for most flight speeds, particularly during the climb. This requirement has led to the development of the variable pitch airscrew in which the blade angle can be varied manually or, more usually, automatically.

A. Betz showed† in the case of a propeller of a finite number of blades that the minimum energy loss, and hence highest efficiency, occurs

† *Handbuch der Physik*, vol. 7, p. 256: 1927.

when the vortex sheets (Figure 109) form a rigid screw surface of constant pitch and move back with a constant velocity along the axis of the propeller and relative to the surrounding air. The flow about these vortex sheets is intimately connected with the circulation about the blades and, hence, the thrust of the propeller. If the jump of potential across the sheets can be calculated, and since the calculation depends on the number of the blades and spacing of the sheets, the thrusts of propellers with various numbers of blades can be compared. The problem is a formidable one mathematically and the first solution due to L. Prandtl, in an Appendix to Betz's Göttingen paper, was obtained by reducing the problem to a two-dimensional one in which the curved vortex sheets were replaced by a series of infinite straight sheets. The results showed that the propeller with a finite number of blades was less efficient than one with an infinite number due to the large radial component of induced velocity occurring near the blade tips. Prandtl represented this loss of efficiency as an effective shortening of the blade radius. The complete three-dimensional solution was given by S. Goldstein.†

Towards the end of the thirties aircraft speeds had become high enough for propeller tip speeds to approach the speed of sound with a corresponding fall of efficiency due to the loss of lift and increase of drag of the outer part of the blade. Designs were produced employing the principle of sweepback and giving blades a scimitar-like shape, but further development was abandoned because of the appearance of the jet engine. Nevertheless, propellers are still under active consideration because of their use in VTOL (vertical take-off and landing) aircraft and helicopters. In the latter case the rotor follows propeller theory in hovering and vertical flight, but in forward flight the rotor is inclined to the airflow, and the theory of the vortex wake is so complicated that it has still to reach a satisfactory state of development.

The inner structure of fluids

Any fluid is the seat of many and varied forms of energy. But what, so to speak, is the inner source of these energy manifestations? For example, why is it that the density of a gas depends on the external pressure applied to it?

† *On the vortex theory of screw propellers*, Roy. Soc. Proc. (A), 123, 440: 1929.

Let us recall first some of the basic concepts worked out at the beginning of this book. By fluid is meant a state of matter in which only a uniform isotropic† pressure can be maintained without indefinite distortion. A *perfect fluid* is frictionless and therefore offers no resistance to flow except through inertial reaction. A homogeneous fluid has the same structure and properties everywhere. An isotropic fluid has local properties independent of rotation of the axis of reference along which those properties are measured. An incompressible fluid is one whose density is substantially unaffected by change of pressure. An elastic fluid is a fluid for which elastic stresses and hydrostatic pressures are large compared with viscous stresses. A Newtonian fluid is a viscous fluid in which the viscous stresses are a multiple of the rate of strain, and the contact of proportionality is the fluid viscosity. A Maxwellian fluid is a viscous fluid in which the stress-strain relationship includes the relaxation effect of the elastic stresses set up by a sudden deformation. A thixotropic fluid is one whose viscosity is a function not only of the shearing stress, but also of the previous history of motion within the fluid.‡

All these classes of fluids possess the property of fluidity, the ability to flow. It is a measure of the rate at which a fluid is deformed by a shearing stress, and is mathematically the reciprocal of the viscosity. The unit of fluidity is the inverse 'poise', the latter being a unit of coefficient of viscosity defined as the tangential force per unit area (dynes/cm^2) required to maintain unit difference in velocity (1 cm/sec) between two parallel planes separated by 1 cm of fluid (1 poise = 1 dyne sec/cm^2 = 1 gm/cm.sec).

Not so long ago, fluidity was thought to be nothing more than the rolling of minute fluid spheres over and between each other. Then man became aware that matter was ultimately composed of atoms, of which there were over ninety different kinds – the chemical elements. The atoms had an electric structure of their own, which does not concern us here, and the heaviest of them (Uranium) was two hundred and thirty-eight times heavier than the lightest (Hydrogen); from 180 to 42,000 million million millions of them went to the grain avoirdupois. Yet man realized that it was mostly from units consisting of two or more atoms that the properties of matter were derived.§ He also discovered that forces of attraction between one atom and another, especially that which became known as chemical affinity, were very powerful under terrestial conditions, and invested all but very few kinds with a horror of loneliness; and that the overwhelming majority of atoms were gathered into clusters of two kinds, known as the molecule and the crystal.

† *Isos* = similar, equal + *tropos* = way of behaviour.

‡ *Scientific Encyclopedia*, fourth edition. Van Nostrand.

§ See, for instance, *The Atom* (Fifth Edition). Sir George Thomson, Oxford University Press, London: 1957.

Two fundamental properties of the atom itself must now be introduced. The first is the interatomic attraction, mainly in the form of chemical affinity, and the second is the heat motion. This latter is due to a ceaseless movement of atoms. A solitary atom can be imagined hurtling along in a straight line with a velocity of furlongs per second, until it meets an obstruction, a wall or another flying particle, when it either sticks (adhesion, cohesion, chemical combination) or continues moving in another direction. Its velocity is lowered by cold and increases with the temperature. When atoms combine, the resulting molecules continue darting about in the same way, but with diminished velocities corresponding to their greater weight. Now, here it is evident that we have an essentially disruptive force, counteracting atomic or molecular attraction: rolling stones gather no moss, as we know. The whole world of matter may be interpreted by the interplay of these two opposed forces, and the broad outcome is that, on the whole, the lower the temperature, the more closely knit is matter.

The most striking effect of cohesion lies in its dividing all matter into three states of aggregation, solid, liquid and gaseous. Ice is a solid body, but it can be melted into water, while the latter can be vaporized. At and above boiling point (100°C = 212°F), the heat motion of water molecules is vigorous enough to prevent them from coalescing. In the liquid interval between 100°C and freezing point (0°C = 32°F) cohesion keeps the molecules in a closely packed mass while still allowing heat-motion to drive them about in a slow, irregular motion. Below freezing point, cohesion gets the upper hand still more: the molecules or atoms are arranged in a rigid, permanent pattern, and each of them can have only enough motion left to execute slight but unceasing vibrations. This latter state continues down to the absolute zero of temperature (−273°C = −459°F), when all motion ceases to exist.

I began this section by saying that a fluid is the seat of varied forms of energy. The idea that heat is one of the forms of manifestation of internal energy, was mooted long before being worked out into a coherent quantitative system, the kinetic theory of gases, mainly by Maxwell in England, Clausius in Germany, and Lomonossov in Russia. They all came to one and the same generalization that wherever there is gas, there is also heat energy. And the presence of gases everywhere around us is evident from countless examples. We create aeroplanes and balloons which fly because of the existence of the atmosphere. When we run fast, we feel the resistance of the air. We are familiar with winds. We infer the restless movement of the particles from the phenomenon of diffusion: a domestic gas leak reaches our consciousness sooner than would be accounted for by draughts. And so on, and so forth.

But how much gas is there around us? I have already discussed this question, therefore I shall make only one or two additional remarks. To use Dr W. A. Caspari's expression, 'an empty half-pint tumbler on your table contains matter (nitrogen and oxygen molecules, about eight thousand million million millions of them) which itself would occupy only about seven-thousandths of a cubic inch'. This may be a very impressive figure. But the fact is that the molecules flow easily, which proves that, in actual fact, they meet very little matter.

When we want to apply pressure to a gas, we have to enclose it (the gas) in some container and provide a sliding piston or other device; and when the pressure is applied, it propagates at once to the whole enclosed gas. When we pump air into a vehicle tyre, we certainly feel a definite resistance to our mechanical effort. The explanation in both cases is one and the same: we reduce the intermolecular distances. With each compressive move of the piston, the molecules butt against the walls, and the sum of their impacts represents what we call 'pressure'. The deeper the piston moves, the greater the pressure. And so the volume of a given quantity of any gas appears to be inversely proportional to the pressure which exists in it (Boyle's law, 1662).

We are now ready to make the final step in our analysis. An average atmospheric pressure at sea-level has been adopted as standard of pressure, and is equal to 760 mm of mercury, or 1·03 kilograms per square centimetre, or 14·72 lb per square inch. If the cross-sectional area of a bicycle pump were 1 square inch, we should have to put only about 15 lb on the piston to send it halfway down. If the pump were full of water, the same weight would move the piston only one-twenty-thousandth of the way down.† And this is so because water and air have vastly different compressibilities; the intermolecular distances of water are very much smaller than those in gases. The compressibility due to the loose structure of gases is relatively enormous.

Thus, when we discuss the volume, weight, density and other properties of a gas, we must ask ourselves, first and foremost: what is its pressure? or, which is nearly the same thing: what is its temperature? For, indeed, pressure in a gas is determined by $p = {}^2/_3(E/V)$, where V is unit volume and E is the translational energy of the molecules, $E = {}^3/_2RT$, and R and T the universal gas constant and temperature (abs.), respectively.

† *The Structure and Properties of Matter*. W. A. Caspari, Ernest Benn, Ltd, London: 1928. For more information on the subject, especially on its early history, we recommend, very strongly, the unique book *On the Nature of Things* by Lucretius (George Bell and Sons, London: 1884).

The velocity of sound

Galileo was the first to state that the denser the fluid, the greater its fluidmechanic resistance. Christian Huyghens, Lazare Carnot, Rene Descartes (1596–1650),† Du Buat, and others, developed this obvious fact of nature into the broader concept: that one and the same fluid can offer different fluidmechanic resistance to one and the same body moving in it. But the conclusion lacked clarity, it remained for a long time a more or less abstract conclusion; if its creators thought in terms of d'Alembert's 'stagnation region', they were on the right track; but there is no evidence that they did. We know, however, that Cauchy and Laplace emphasized that their equations were valid for incompressible flows: would it be too risky to deduce from here that they were also aware of the 'compressibility' effects? We have shown earlier that Euler and Lagrange, especially the latter, tried to work out, not entirely unsuccessfully, mathematical relationships between ρ and p: would it be too much to say that they, too, knew of the significance of compressibility?

There were many people who, for one reason or another, tried to work on the problem. But one man played here the role of the integrator of the previous bits and pieces, so to speak. And he was Pierre Henry Hugoniot (1851–87), the outstanding French ballistician.

Helmholtz's theorems do not hold if the forces acting upon the fluid have no potential, or the mass density ρ is not a function of the pressure p alone, because the flow will, or may, have discontinuities. But a discontinuity is stationary if its surface, though moving about in space and changing its form, is composed always of the same fluid particles, if, mathematically speaking, the equation of the surface does not contain time. When it does contain time, then the discontinuity is not attached to the same particles of the fluid. In this last case, we shall say that the discontinuity constitutes a wave, and that it is propagated in the fluid (Felix Savart showed that sound waves are propagated in water in the same way as in solids).

Jacques Salomon Hadamart (1865–1912), a French mathematician, showed‡ that for a stationary discontinuity the normal component of velocity remains continuous across the wave – this is, by the way, the

† *Lettres de Mr. Descartes*. Par la Compagnie des Libraires, Paris, M.DCC.XXIV, 6 volumes.

‡ *Leçons sur la propagation des ondes et les équations de l'hydrodynamique*, Paris: 1903.

foundation stone of today's theory of shockwaves. And Hadamart's theorem proved to be in full agreement with an earlier concept by Hugoniot (called 'acceleration wave'), which allowed mathematical equations to be written across fluidmechanic discontinuities. The analysis and solution of these equations resulted in the following fundamental theorem of Hugoniot:†

In a non-viscous compressible fluid there are only two kinds of discontinuities possible: longitudinal, which are propagated with the velocity $\sqrt{dp/d\rho}$, and transversal, which are not propagated at all and which always affect the same particles. The first of these are waves, and the second are stationary discontinuities.

This is what Hugoniot wrote about Laplace's formula $\sqrt{dp/d\rho}$ for the speed of sound (to take only one paragraph from his analysis):

'Having established the equation of motion of a perfect gas in an adiabatically isolated cylinder, we obtain the velocity of propagation from the equation

$$\left(\frac{d\xi}{dt}\right)^2 = \frac{\gamma p}{\rho}\left(1 + \frac{\partial u}{\partial x}\right)^{-(\gamma+1)}$$

When the gas is at rest, $u = 0$ is the solution for which $\partial u/\partial t = 0$, therefore

$$\left(\frac{d\xi}{dt}\right)^2 = a^2 = \frac{\gamma p}{\rho}, \text{ or } a = \sqrt{\gamma p/\rho},$$

an expression which has been known since Laplace, but which has never been rigorously demonstrated.'

Clear enough. And the same law can be derived for spherical waves.

G. F. B. Riemann was, however, the first who tried to calculate the relations between the states of gas before and after the shock wave, but made a mistake, later corrected by W. J. M. Rankine, the British engineer, and then by Hugoniot. Riemann thought that the change across a shock would be isentropic, and hence that the entropy also would remain unchanged, which is wrong, of course.

I think that the name 'velocity of sound' is somewhat misleading. It would, perhaps, be more logical to call it propagation of small disturbances, or of small pressure changes; for that is precisely what it represents, and what makes sound. In this sense, a sound wave consists of compressional and rarefactional displacements of fluid particles parallel to the direction of advance of the disturbance; longitudinal waves of small-amplitude adiabatic oscillations.

† *Mémoire sur la propagation du mouvement dans les corps et spécialement dans les gases parfaits*. H. Hugoniot, Journal de l'Ecole Politechnique, Paris: 1887–9.

We associate these phenomena with compressible fluids only, because in the incompressible fluid the small pressure or density changes propagate at an infinite velocity, thus reaching the whole fluid instantaneously, and the question of acoustic waves does not arise; while in the compressible fluid they propagate with a finite velocity – sonic velocity – and play an important role in the formation of the flow structure of the fluid.

Ernst Mach (1838-1916) and others

Ernst Mach, an Austrian physicist and ballistician, who became a philosopher, was one of the creators of supersonic aerodynamics.† We shall see presently that this branch of Fluidmechanics studies, among other things, the so-called *shock waves*. To make these visible, Mach used the schlieren method, an optical system of observation of density changes, invented by August Joseph Ignaz Töpler (1836–1912), a German physicist who worked in the field of acoustics.‡ (Figure 113.)

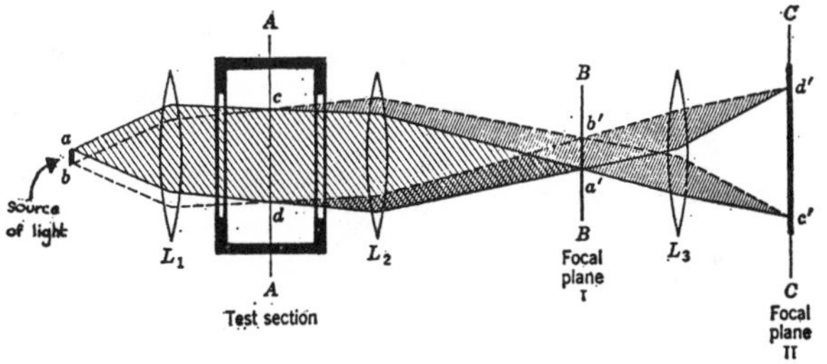

Fig. 113. A schlieren arrangement

Jacob Ackeret, another major creator of gasdynamics, who discovered that the ratio of the speed of fluid flow (v) to the speed of sound (a) in that flow was an important dimensionless instrument of gasdynamic analysis, thought that it might be a good idea to name this ratio after Ernst Mach, since he (Mach) was the first to show that compressibility effects in gas flows depend not on v as such, but on its ratio to a. And so the Mach number was born, $M = v/a$; it shows how much greater or smaller is the velocity of a flow than the velocity of sound in that flow.

† *Ballistisch Photographische Versuche, Sitzungberichte der Wiener Akademie der Wissenschaften*, 98. E. Mach and L. Mach: 1889. *Optische Untersuchung der Luftstrahlen*, ibid., 98. E. Mach and P. Salcher: 1889. We should also mention his other famous book, *The Science of Mechanics*, a critical and historical account of the development of the subject. The English translation (The Open Court Publishing Company, La Salle, Illinois: 1974) has an excellent introduction by Karl Menger.

‡ *Beobachtungen nach einer neuen optischen Methode.* A. Töpler, Bonn: 1864.

When $M < 1$, $v < a$, therefore, the flow is called *subsonic*. When $M = 1$, $v = a$, the flow is transsonic. When $M > 1$, $v > a$, the flow is supersonic. When $M \geqslant 5$, $v \geqslant 5a$, the flows are called *hypersonic*. These are, however, somewhat artificial boundaries. For example, the term transsonic is sometimes used to describe a region of Mach numbers in which $(1 - M)$ changes its sign, or a region of $0{\cdot}82 \leqslant M \leqslant 1{\cdot}2$, approximately.

You need not be an aerodynamicist to carry out a simple experiment in a pond, in your own bathroom, or in any vessel filled with water. By squeezing a wet sponge, let drops fall on the water surface, at one- or two-second intervals, say. If they fall at one and the same point, you will observe concentric wave circles spreading out, the centre of disturbance (the point of impact of the drops) remaining at the same spot (Figure 114a). Now let the sponge, and consequently the point of disturbance,

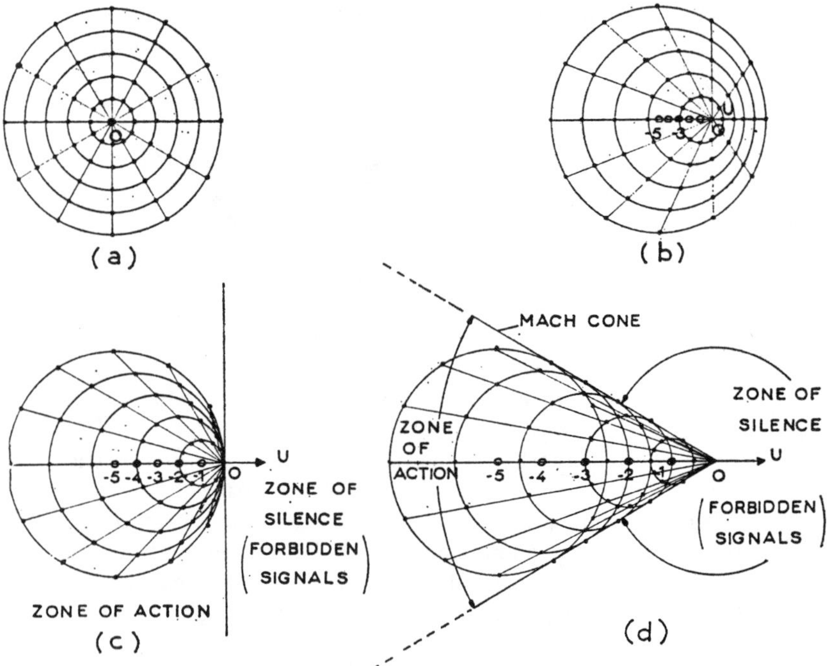

Fig. 114. Stages of formation of the Mach cone:
(a) Sound waves exist everywhere around the point of disturbance
(b) The front zone of sound has become 'narrower'
(c) Sound exists only to the left of the vertical line
(d) All the sound waves are inside the Mach cone

move from left to right (say): you will then see that in front of the point of disturbance the wave circles become condensed and behind it – rarefied (Figure 114b). One could say that the point of disturbance runs after the disturbances ahead of it, and runs away from the disturbances behind it. When M $\approx$ 1, the source catches up with the circles ahead and we have Figure 114c. When M $>$ 1, the source moves faster than the propagation of the disturbances it creates (Figure 114d).

If you could measure the strength of any of the circular waves, you would find that they are very weak waves. When a fluid particle crosses such a wave, it experiences very small and, for practical purposes, negligible changes in its density, pressure, temperature, momentum, etc. And the same is true, of course, in the atmosphere, in any gas.

So, if you imagine the above pictures in three dimensions, in the atmosphere, the following conclusions emerge: in Figure 114a sound waves exist everywhere around the point of disturbance, i.e. there is no zone of silence; in Figure 114b the front zone of sound has become 'narrower' and the rear zone 'broader', which means that the intensity of sound is stronger in the first and weaker in the second zone (which is why the whistle of a train locomotive is stronger when it approaches and gets weaker when it passes you); in Fig. 114c sound exists only to the left of the vertical line, i.e. the zone to the right of that line has no sound waves, and you hear nothing; in Fig. 114d all the sound waves are inside the cone, the Mach cone, and therefore the whole exterior zone is free of sound.

Let us recall our earlier remarks. Sir Isaac Newton was the first to compute the speed of sound on the basis of the elasticity and density of air which were established before him. From the well-known Robert Boyle's law (1662), also known as Mariotte's law, expressing the isothermal pressure-volume relation for a body of ideal gas (at constant temperature) in the form $pv =$ const, Newton deduced that the speed of sound $a = \sqrt{p/\rho} = 916$ ft/sec $= 280$ m/sec. But experiments proved that the actual a was 20% higher. Laplace explained why the compression and rarefaction in a sound wave follow Poisson's adiabatic curve, i.e. why they are adiabatic. Heating in the compression and cooling in the rarefaction waves strengthen the pressure changes in them, thereby increasing the sonic speed according to the formula $a^2 = \gamma p/\rho$, where $\gamma = C_p/C_v$.

From Figure 114d we have $\sin\alpha = at/vt = a/v = 1/(v/a) = 1/M$; from which it is obvious that the vortex angle of the Mach cone would be different for Newton's and Laplace's values of a; therefore the consequences of the cone would be different – serious mistakes in the whole edifice of gas-dynamics in the first case, and accurate results in the second case. This shows once again the greatness of Laplace's role in the history of Fluidmechanics.

The surface of the cone consists of Mach lines, or Mach disturbances. Like the waves created in your bath, they are very weak lines, in the sense that air particles crossing them experience infinitesimal changes of pressure, temperature, density, velocity. We thus have a deep analogy between the phenomena observed on the free surface of water in your bath and in gasdynamics (this analogy was discussed and formulated by L. Prandtl, 1928).

One question arises at once: what happens to the inner physical state of fluid inside the wave? This is, really, one of the most fundamental questions of compressible Fluidmechanics, first put forward by Earnshaw in connection with the problem of the maximum discharge velocity.† About 17 years later, the question was analysed also by Hugoniot.

The following two (deliberately simplified) examples lead to the answer as it is known today. Assume that when a gas particle moving along the central stream-line hits the stagnation point O (Figure 115), it

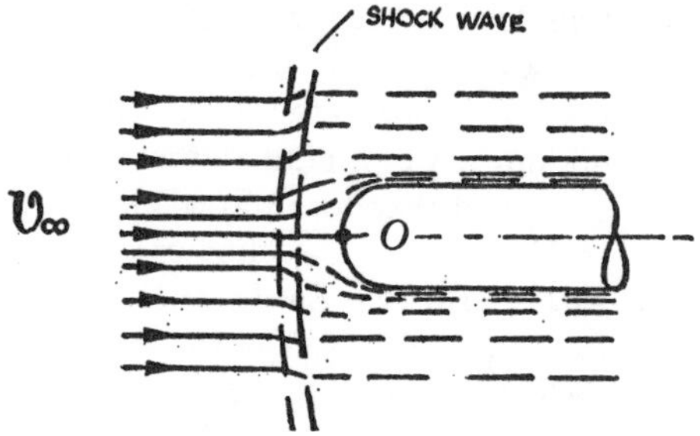

Fig. 115. The formation of a shock wave

loses its kinetic energy $\rho v^2/2$ and stops. The particle behind hits it; the same happens to millions and millions of all the other particles. So, around the point O we have a zone of pile-up of particles, where pressure is high and the lost kinetic energy becomes heat energy.

There are only two ways of escape for the particles: upwards and downwards. The escape takes place under the stagnation region pressure, therefore particles 'shoot out' of the region, each carrying a certain momentum. But while moving in the transverse direction, they are hit by the oncoming particles of the flow (above and below the central line, thus themselves becoming sources of disturbance) and are compelled to change their direction. So, instead of being a straight line perpendicular

† *Mathematical Theory of Sound.* Rev. Samuel Earnshaw, Phil. Trans., 150, 133: 1858.

to the flow, the shock wave becomes curved. Immediately before the body it remains, however, normal to the flow, therefore the central parts of shock waves of this type are said to be *normal shock waves*. The higher the speed of the flow, that is, the higher its Mach number, the more twisted (in the downstream direction) is the wave, and the smaller its stand-off distance.

For the second example, consider now Figure 116. A particle moving

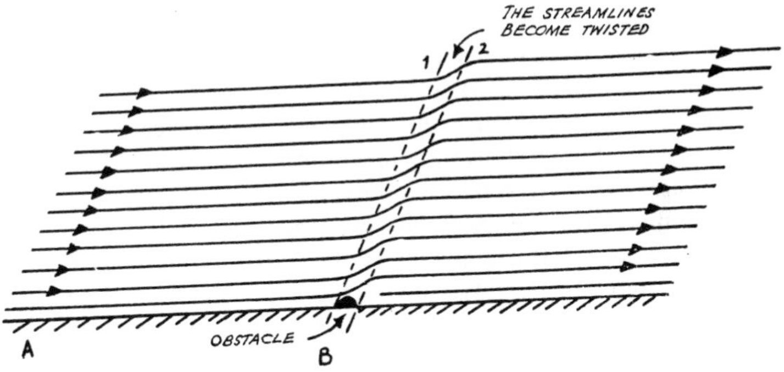

Fig. 116. A scheme of shock wave formation on a wall

along the solid surface *AB* hits some obstacle at *B*, bounces upwards, hitting, in turn, its neighbour immediately above; the latter hits its neighbour, and so on. All these disturbed particles tend to move upwards. But the oncoming particles (i.e. flow) prevent them from continuing their upward motion, they lose parts of their transverse momentum very quickly, and the stream-lines become twisted within a strip between lines 1 and 2, as can be seen also in Plate 19.

In both cases, the consequence is a shock wave. The wave is very much stronger than the sonic waves discussed earlier. It stands there as long as the 'shock conditions' of the flow are maintained, therefore the flow particles have to go through it. And when they cross it, their kinematic and thermodynamic characteristics experience a rather sharp change, a shock. As to the physical state of the fluid inside the wave, two limiting cases are of particular importance: (a) the case of very small viscosity, and (2) the case of small heat conductivity. These cases were studied by E. Jouquet,† Rayleigh,‡ R. Becker,§ L. Prandtl|| and others. A very full, probably the best, mathematical treatment of the whole

† Journ. de Math., 6, 6: 1904.
‡ Proc. Roy. Soc., 84, 247: 1910.
§ ZS für Physik, 8, 326: 1920.
|| ZS f.d. gesamte Turbinenwesen: 1906.

theory of shock waves appears in the famous work by N. E. Kotchin, I. A. Kibel, and N. V. Rose.†

The Chaplygin-Khristianovich method

Although the main stream may still be subsonic ($M_0 < 1$), at some point, or points, of the surface of a body moving in a gas, the local flow speed must be supersonic ($M < 1$), if shock wave or waves exist. The free stream Mach number at which this happens (i.e. the local flow velocity becomes equal to the local velocity of sound) is called the critical Mach number, M_{cr}. The corresponding pressure is called critical pressure, p_{cr}, or minimum pressure, p_{min}, because when $M_0 = M_{cr}$, or $v = v_{cr}$, we have $p = p_{min}$. Khristianovich of the USSR showed that, for an aerofoil, M_{cr} and the minimal pressure coefficient $\bar{p}_{min}$ (which is defined as $2\triangle p/\rho v^2$) depend on each other as shown in Figure 117.

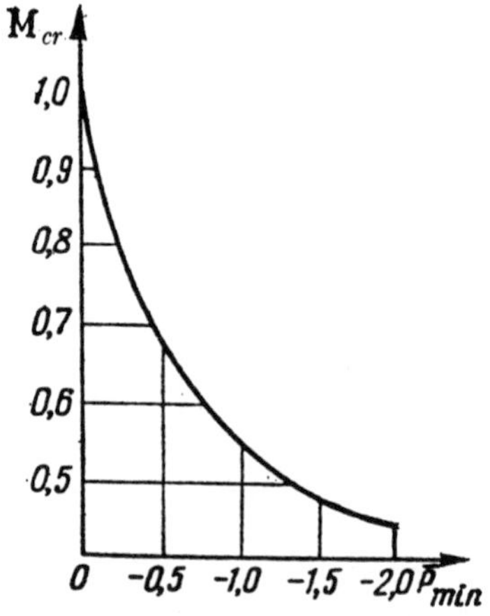

Fig. 117. Khristianovich's diagram $p_{min} = f(M)$

The notion 'critical' has only one meaning: no shock wave can occur below the critical Mach number. The other interesting fact is that shock wave formation at low Reynolds numbers is considerably different from that at high Reynolds numbers (at which the flow is at least partly

† Teoretitcheskaya Gidromekhanika (Theoretical Hydromechanics), vol. 2, 727 pp. Moscow: 1963 (in Russian).

turbulent). When the boundary layer is turbulent, it has a higher resistance to the so-called 'shock stall' – the flow separation. Such a separation is usually caused by the pressure rise due to the shock: a very complex matter, first investigated by Jacob Ackeret.

The theory of high *subsonic* flows with compressibility effects was studied, intensively and successfully, by Sergei Alekseyevich Chaplygin, or Tchaplygin (1869–1942) in the CAGI* of the USSR. His contributions to fluidmechanics were numerous; his work dealing with high subsonic flows was historic.† In it, he introduced new independent variables and transformed the general equations of gasdynamics into equations in the hodograph plane, and linearized them in this. They are:

$$\frac{\partial\varphi}{\partial\tau} = -\frac{1}{2\tau}\frac{1-\dfrac{\gamma+1}{\gamma-1}\tau}{(1-\tau)^{\gamma/(\gamma-1)}}\frac{\partial\psi}{\partial\theta}$$

$$\frac{\partial\varphi}{\partial\theta} = \frac{2\tau}{(1-\tau)^{1/(\gamma-1)}}\frac{\partial\psi}{\partial\tau}$$

where $\tau = (\gamma-1)\lambda^2/(\gamma+1) = (v/v_{max})^2$, $v_{max} = \sqrt{(\gamma+1)/(\gamma-1)}a_{cr}$, φ is the velocity potential, ψ the stream function, θ the angle between v and ox-axis, $\psi = C_p/C_v$, $\lambda = v/a_{cr}$.

Another CAGI aerodynamicist, Sergei Alekseyevich Khristianovich (born in 1908),‡ a close collaborator of Tchaplygin, introduced another variable s, such that

$$ds = \sqrt{\frac{1-\lambda^2}{1-\dfrac{\gamma-1}{\gamma+1}\lambda^2}}\,\frac{d\lambda}{\lambda},$$

and reduced Tchaplygin's equations to the form

$$\boxed{\begin{aligned}\frac{\partial\varphi}{\partial\theta} &= \sqrt{K}\,\frac{\partial\psi}{\partial s}\\ \frac{\partial\varphi}{\partial s} &= -\sqrt{K}\,\frac{\partial\psi}{\partial\theta}\end{aligned}},$$

* Central'nyi Aero-Gidrodinamitcheskii Institut (Central Aero-Hydrodynamic Institute), near Moscow, the second-largest such establishment in the world.

† On Gas Streams (in Russian: O gazovykh struyakh), Moscow University: 1904. Also in his Coll. Works, vol. II, Moscow: 1938. Available in English, NACA TM No. 1063, 1944.

‡ Trudy GAGI, issue 481, Moscow: 1940.

where

$$K = \frac{1 - \lambda^2}{\left(1 - \dfrac{\gamma - 1}{\gamma + 1}\lambda^2\right)^{(\gamma+1)/(\gamma-1)}}$$

Historical objectivity demands the statement here that the Khristianovich form of Tchaplygin's equations was established before him by N. A. Slezkin† and L. S. Leibenson.‡ But Khristianovich carried out their further study and developed a method which became (deservedly) known as 'Khristianovich's Method'. However, strangely enough, it does

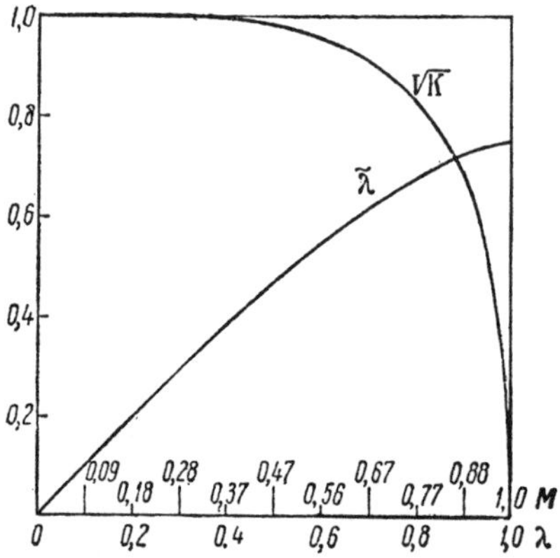

Fig. 118. Khristianovich's diagram for $\sqrt{K} = f(M)$

not appear in Western textbooks, and the average Western aerodynamicist remains unfamiliar with it.

When λ is not too close to 1, the function $\sqrt{K(\lambda)} \approx 1$, as can be seen from Figure 118 and from the following table (for $\gamma = 1{\cdot}4$):

λ	M	$\sqrt{K}$	λ	M	$\sqrt{K}$	λ	M	$\sqrt{K}$
0·00	0·0000	1·0000	0·35	0·3228	0·9965	0·70	0·6668	0·9221
0·05	0·0457	1·0000	0·40	0·3701	0·9940	0·75	0·7192	0·8925
0·10	0·0913	1·0000	0·45	0·4179	0·9899	0·80	0·7727	0·8416
0·15	0·1372	0·9999	0·50	0·4663	0·9840	0·85	0·8274	0·7740
0·20	0·1832	0·9996	0·55	0·5152	0·9754	0·90	0·8834	0·6788
0·25	0·2294	0·9991	0·60	0·5649	0·9632	0·95	0·9409	0·5092
0·30	0·2759	0·9982	0·65	0·6154	0·9461	1·00	1·0000	0·000

† 'Doklady' of the Academy of Sciences of the USSR, vol. 3, No. 9: 1936.
‡ 'Doklady' of the Academy of Sciences of the USSR, No. 9: 1935.

The values of $\sqrt{K}$ are easily computed from

$$\sqrt{K} = (1 - M^2)^{\frac{1}{2}}\left(1 + \frac{\gamma - 1}{2} M^2\right)^{1/(\gamma-1)}$$

Thus, for subsonic flows we can put $\sqrt{K} \approx 1$ and, consequently, the equations assume the form

$$\frac{\partial\varphi}{\partial\theta} = \frac{\partial\psi}{\partial s}, \frac{\partial\varphi}{\partial s} = -\frac{\partial\psi}{\partial\theta},$$

which are analogous to the Cauchy-Riemann equations for incompressible flows,

$$\frac{\partial\varphi}{\partial\tilde{\theta}} = \frac{\partial\tilde{\psi}}{\partial\tilde{s}}, \frac{\partial\tilde{\varphi}}{\partial\tilde{s}} = -\frac{\partial\tilde{\psi}}{\partial\tilde{\theta}}$$

in the $(\tilde{s}, \tilde{\theta})$ plane ($\sim$ indicates incompressible flow). Obviously, also $d\tilde{s} = d\tilde{\lambda}/\tilde{\lambda}$. Khristianovich then puts

$$ds = \sqrt{\frac{1 - \lambda^2}{1 - \frac{\gamma - 1}{\gamma + 1}\lambda^2}}\frac{d\lambda}{\lambda} = d\tilde{s} = \frac{d\tilde{\lambda}}{\tilde{\lambda}}$$

and integrates it. The constant of the integration is determined for the condition $lim(\lambda/\tilde{\lambda})$ when $\lambda \to 0$. The actual integration is far too complex for this book, therefore I omit it and give instead the following table of its numerical results, which should be compared with Figure 118:

λ	M	$\tilde{\lambda}$	λ	M	$\tilde{\lambda}$	λ	M	$\tilde{\lambda}$
0·00	0·00	0·00	0·035	0·3228	0·3410	0·70	0·6668	0·6251
0·05	0·0457	0·0500	0·40	0·3701	0·3862	0·75	0·7192	0·6568
0·10	0·0913	0·0998	0·45	0·4179	0·4307	0·80	0·7727	0·6857
0·15	0·1372	0·1493	0·50	0·4663	0·4734	0·85	0·8274	0·7110
0·20	0·1832	0·1983	0·55	0·5152	0·5144	0·90	0·8834	0·7324
0·25	0·2294	0·2467	0·60	0·5649	0·5535	0·95	0·9409	0·7483
0·30	0·2759	0·2943	0·65	0·6154	0·5904	1·00	1·0000	0·7577

Thus, there exists a definite connection between λ and $\tilde{\lambda}$, which is independent of both the configuration of the body and the character of the flow. And if so, the compressibility effects can be determined in a straightforward manner from the well-known isentropic formulae,

$$\frac{p}{p_0} = \left(\frac{1 - \frac{\gamma-1}{\gamma+1}\lambda^2}{1 - \frac{\gamma-1}{\gamma+1}\lambda_0^2}\right)^{\gamma/(\gamma-1)}$$

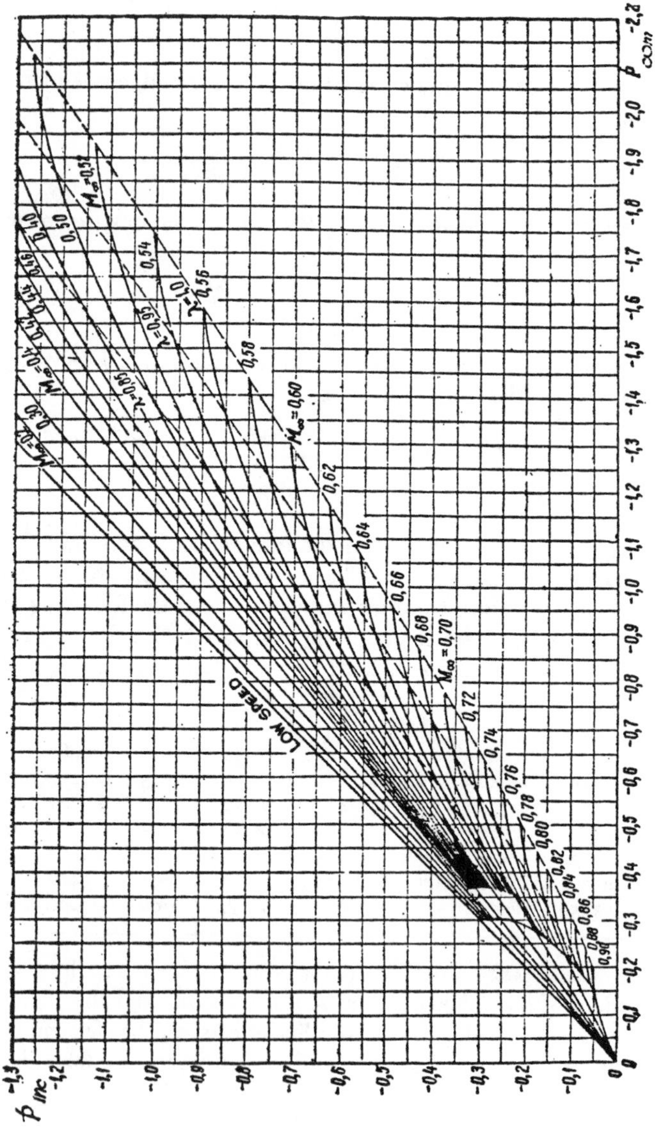

Fig. 119 and (opposite) 120. Khristianovich's charts for the determination of high subsonic compressibility effects at various free stream Mach numbers

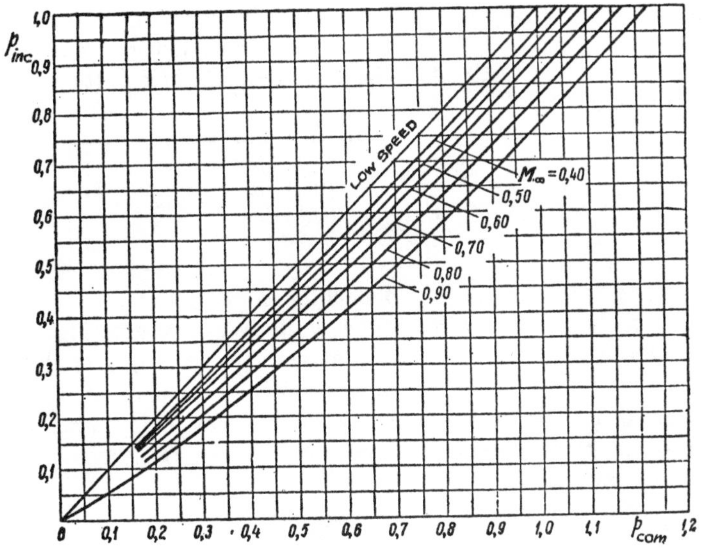

and

$$\frac{a_0^2}{a_{cr}^2} = \frac{1 - \dfrac{\gamma-1}{\gamma+1}\lambda_0^2}{1 - \dfrac{\gamma-1}{\gamma+1}\lambda_{cr}^2} = \frac{\gamma+1}{2}\left(1 - \frac{\gamma-1}{\gamma+1}\lambda_0^2\right),$$

where, obviously, $\lambda_{cr}^2 = (v_{cr}/a_{cr})^2 = 1$. Then,

$$\bar{p}_{com} = \frac{2(p-p_0)}{\rho_0 v_0^2} = \frac{2p_0}{\rho_0 v_0^2}\left[\left(\frac{1-\dfrac{\gamma-1}{\gamma+1}\lambda^2}{1-\dfrac{\gamma-1}{\gamma+1}\lambda_0^2}\right)^{\gamma/(\gamma-1)} - 1\right] =$$

$$= \frac{\gamma+1}{2}\,\frac{1-\dfrac{\gamma-1}{\gamma+1}\lambda_0^2}{\lambda_0^2}\left[\left(\frac{1-\dfrac{\gamma-1}{\gamma+1}\lambda^2}{1-\dfrac{\gamma-1}{\gamma+1}\lambda_0^2}\right)^{\gamma/(\gamma-1)} - 1\right]$$

Obviously, also

$$\bar{p}_{inc} = 1 - (\tilde{\lambda}/\tilde{\lambda}_0)^2,$$

subscripts 'com' and 'inc' standing for 'compressible' and 'incompressible'. The table and these two formulae solve the problem completely. Figures 119 and 120 give the relationships between $\bar{p}_{com}$ and $\bar{p}_{inc}$ at various subsonic Mach numbers.

Such, then, is Kristianovich's method. Its advantage is its applicability to a wide range of body configurations. But there are also other methods. For example, S. A. Tchaplygin developed single (but approximate) relationships,†

$$\bar{p}_{com} = -\frac{2\sqrt{1+\mu_0^2}}{\mu_0^2}\left(\sqrt{1+\mu^2} - \sqrt{1+\mu_0^2}\right)$$
$$\bar{p}_{inc} = 1 - (\tilde{\mu}/\tilde{\mu}_0)^2$$

where $\mu^2 = v^2/a_0^2$ (in this case 'o' denoting stationary gas).

In 1928, for the 'thin' aerofoil, Glauert derived‡ a relationship between the subsonic free stream Mach number M_0 and $\bar{p}_{com}$ in the form $\bar{p}_{com} = \bar{p}_{inc}\sqrt{1-M_0^2}$, where $\bar{p}_{inc}$ is the pressure coefficient at a speed low enough to consider the flow incompressible. Then, independently from each other, H. S. Tsien§ and Theodore von Karman|| arrived at one and the same formula,

$$\bar{p}_{com} = \frac{\bar{p}_{inc}}{\sqrt{1-M_0^2} + \frac{1}{2}\left(1-\sqrt{1-M_0^2}\right)\bar{p}_{inc}}$$

Finally, the minimum pressure coefficient emerged as a function of the critical Mach number,

$$\bar{p}_{min} = 1 - \frac{1}{M_{cr}^2}\left(\frac{2}{\gamma+1} + \frac{\gamma-1}{\gamma+1}M_{cr}^2\right)^{\gamma/(\gamma-1)}$$

This formula was derived in 1949 by Prof. G. F. Burago of the Zhukovsky Academy of Aeronautics in Moscow.

The drag wall

Towards the end of the thirties, aviation had reached 'the blind wall of resistance'; although the speeds of flight were still subsonic, at Mach numbers of about 0·65–0·72 the aerodynamic resistance – drag – rose fairly suddenly, like a wall against further increase of speed. This was due to the compressibility effects which caused sharp discontinuities in the flow, unstable formation of shockwaves accompanied by trans-

† Reference on page 199.
‡ Proc. Roy. Soc., vol. A118: 1928.
§ Journ. Aero. Sci.: August, 1939.
|| Journ. Aero. Sci.: July, 1941.

formation of a large amount of flow energy into heat and into pressure pulses that consumed a major portion of the available propulsive energy. In short, aviation came up against a new drag – ***wave drag***.

These unpleasant processes, erratic formation of shock waves and unstable mechanism of the flow, exist throughout the whole range of high subsonic and transsonic aerodynamics, as illustrated by Plate 20.† In Plate 23 we have typical normal shockwaves obtained in the Department of Aeronautics, The City University, London.

Apart from Hugoniot and others already mentioned, Professor William John Macquorn Rankine (1820–72), the British molecular physicist, must be credited with the physical and mathematical theory of shock waves. His celebrated work called 'On the Thermodynamic Theory of Waves of Finite Longitudinal Disturbance' was first published in the Philosophical Transactions of the Royal Society, 160, 277–88, 1870, and reappeared in due course many times.‡ Careful reading of this work shows his complete physical and mathematical understanding of the subject.

The object of Rankine's investigation was to determine the relations which must exist between the laws of the elasticity of any substance, whether gaseous, liquid or solid, and those of the wave-like propagation of a finite longitudinal disturbance in that substance. In other words, of a disturbance consisting in displacements of particles along the direction of propagation, the velocity of displacement of the particles being so great that it was not to be neglected in comparison with the velocity propagation. In particular, the investigation aimed at ascertaining what conditions as to the transfer of heat from particle to particle must be fulfilled in order that a finite longitudinal disturbance may be propagated along a prismatic or cylindrical mass without loss of energy or change of type: the word 'type' being used to denote the relation between the extent of disturbance at a given instant of a set of particles, at their respective undisturbed positions.

Then there was the outstanding work 'Ueber einige Bewegungen eines Gases bei Annahme eines Geschwindigkeitspotential', by P. Molenbroek, first published in the *Archiv Math. Phys.*, (2), 9, 157, 1890, which enriched Gasdynamics with valuable mathematical techniques. And the famous 'Ueber zweidimensionale Bewegungsvorgänge in einem Gas, das mit Ueberschallgeschwindigkeit strömt' – the historic Inaugural Dissertation zur Erlangung der Doktorwürde, by Theodor Meyer, first

† *Wind-Tunnel Technique*. R. C. Pankhurst and D. W. Holder, Pitman & Sons, London: 1952.

‡ See for example, *Foundations of High Speed Aerodynamics*. Compiled by Robert T. Beyer, Dover Publications, Inc., New York: 1951.

published in 1908 in Göttingen, and then reprinted partly or fully many times.

Less known are, however, the important contributions made by Cranz alone or in co-operation with someone else. He tried very hard, probably harder than anyone else before the First World War, to study such problems as the relationship between the speed of flight and drag of a bullet, the behaviour of gas in front of a fast moving body, the role of the head-configuration of a body in the formation of drag, etc.

Cranz was, of course, familiar with Hugoniot's and Rankine's theories, therefore we can safely say that he also studied shock waves, as will be obvious a little later on. But it is not clear from his writings whether he was aware of the critical interdependence of $\bar{p}_{min}$ and emergence of shock waves. Nor can we say that he distinguished between high-subsonic and trans-sonic gasdynamics. At any rate, he produced no solutions of the kind we discussed a little earlier; in fact, one could go as far as to say that Cranz was unsuccessful in his attempt to explain the transsonic 'drag wall'. But Col. Haupt, an eminent German artillerist, who was his contemporary and friend, thought that he knew the answer. In several articles,† P. Haupt developed a theory of drag of bullets and artillery shells based upon the kinetic theory of gases at low and high subsonic speeds. Again, not very successful, but the interesting point is that his arguments vibrated around compressibility effects and heating: a good approach, indeed.

Still more interesting, Hauptmann J. Schatte published two articles‡ called 'Die experimentalle Ermittelung der gunstigsten Geschossform auf Grund einer neuen Methode zur Messung des Luftwiderstands', containing excellent schlieren and interferometer pictures of supersonic shock-waves and flow patterns (Plate 22). It was the same Schatte whose work on the optical method of visualization of waves (and flow patterns) created by bullets received an enthusiastic reaction from many physicists, aerodynamicists and military experts (Figure 121).

Then there was a French ballistician called J. Didion, who suggested§ semi-empirical formulae, and tables, for the computation of the drag characteristics of artillery shells of various head configurations. It contains, by the way, references to the so-called 'head angle', which is, in fact, something like a precursor of the sweepback wing.

We then have F. Siacci's (1839–1907) famous work‖ – probably one of the most valuable contributions to the theory of drag before Aero-

† Artilleristische Monatshefte, Nr 40, 1910.
‡ 'Kriegstechnische Zeitschrift', 1913, 1. Heft und 3. Heft.
§ 'Traité de ballistique', Paris: 1848.
‖ Rivista d'artigleria e genio, vol. 1: 1896.

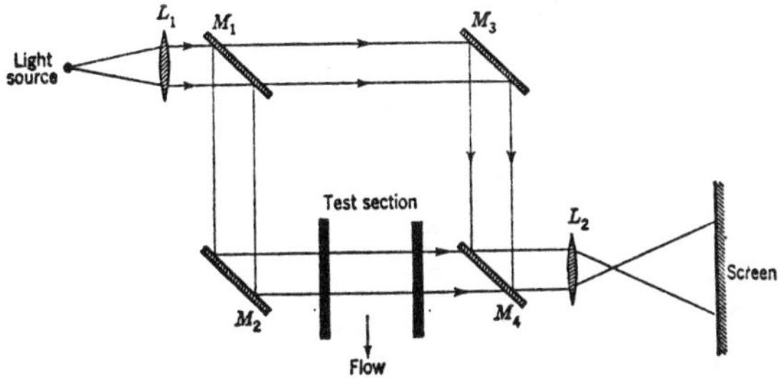

Fig. 121. Schatte's optical system for visualisation of flow

dynamics emerged as a coherent set of laws and theorems. He showed that when the speed of a body moving in the atmosphere approaches the speed of sound, the $R \propto v^2$ law ceases to be valid.

When a 'heavy' material point moves in the atmosphere vertically upward, both its weight W and air resistance R tend to decelerate it and reduce its velocity v. Therefore the equation of the motion should be written in the form

$$\frac{W}{g}\frac{dv}{dt} = -W - R, \text{ whence}$$

$$\frac{dv}{dt} = -g - \frac{gR}{W}$$

Here gR/W has the dimension of acceleration; Siacci denoted it by $I = cf(z)F(v)$; $f(z)$ represents the law of change of density with altitude, so that $f(0) = 1$; $F(v)$ is the Siacci function, for which a special table has been tabulated; c is the ballistic coefficient determined by $c = id^3 \times 10^3/W$, d being the diameter of the shell and i the so-called 'configuration coefficient'.

When the body flies at speeds for which $R \propto v^2$ (i.e. when $v < 200$ m/sec), the ratio $F(v)/v^2 = \text{const}$. But when it moves at speeds $v \geqslant 200$ m/sec (i.e. with compressibility effects), the function changes as the pressure changes gradually from a higher to a lower value through the expansion waves, the supersonic velocity remaining finite, provided that the (convex) change of direction is not too sharp, like Borda's sudden expansion.

The mathematical, kinematic and thermodynamic characteristics of all these and other phenomena can be found in Dr Theodor Meyer's work,† or, in a condensed form, in one of the following outstanding

† *Uber Zweidimensionale Bewegungsvorgange in einem Gas, das mit Ueberschallgeschwindigkeit strömt.* von Dr Th. Meyer, Berlin: 1908.

textbooks: (1) G. N. Abramovich, *Prikladnaya Gazovaya Dinamika* (Applied Gas Dynamics), Moscow, 1953 (in Russian); (2) N. E. Kotchin, I. A. Kibel', N. V. Rose, *Teoretitcheskaya Gidromekhanika* (Theoretical Hydromechanics), Vol. 2, Moscow, 1963 (in Russian, probably one of the best textbooks in Fluidmechanics); (3) L. G. Loytsiansky, *Mekhanika Zhidkosti i Gaza* (The Mechanics of Fluids and Gases), Moscow, 1959 (in Russian, an excellent textbook); (4) L. D. Landau and E. M. Lifshitz, *Mekhanika Sploshnykh Sred* (The Mechanics of Continuous Media), Moscow, 1954 (in Russian); (5) H. W. Liepmann and A. Roshko, *Elements of Gasdynamics*, New York, 1957; (6) K. Oswatitsch, *Gasdynamics*, Wien, 1952; (7) A. M. Kuethe and J. D. Schetzer, *Foundations of Aerodynamics*, New York, 1961; (8) J. H. Dwinnel, *Principles of Aerodynamics*, New York, 1949; (9) D. O. Dommash, S. S. Sherby, Th. F. Connolly, *Airplane Aerodynamics*, New York, 1951 – to name only a few.

The further growth and consolidation of the science of supersonic flows and shock waves was affected by the important contributions of thousands of academics and research workers: obviously, we cannot list them all. But A. Busemann's 'Infinitesimal kegelige Überschallströmung' (1943); L. Prandtl's and A. Busemann's 'Aerodynamischer Auftrieb bei Überschallgeschwindigkeit' (1935); J. Ackeret's 'Gasdynamik' (Handb. der Physik, Bd. VII, 1927, and 'Über Luftkräfte auf Flügel, die mit grosserer als Schallgeschwindigkeit bewegt werden' (1925); Th. von Karman's 'Supersonic Aerodynamics' (1947); S. A. Khristianovich's 'Obtekaniye Tel Gazom Pri Bol'shikh Skorostyakh' (1940), Ya. B. Zel'dovich's 'Theory of Shock Waves' (1946), R. Sauer's 'Einführung in die theoretische Gasdynamik' (1944/5) – and many others – will always be remembered as the pace-making contributions.

There are, however, some aspects of gasdynamics – and here I must return also to transsonic gasdynamics – which were born several decades earlier. In 1910, Hauptman Bensberg and Dr Cranz published an article† containing extremely interesting information, including photographs of flying bullets. In this and subsequent series of articles, we see the Cranz Curves, obtained theoretically and experimentally. They give detailed descriptions of the now famous methods of visualization of flows, Schlieren and Interferometer methods, with all the basic schemes and technical details. Figure 122 shows the now very well-known pattern of change of drag with Mach number (admittedly, Cranz did not use $M = v/a$, he used v). The thoughtful reading of various works by

† Über eine photographische Methode zur Messung von Geschwindigkeiten und Geschwindigkeitsverlusten bei Infanteriegeschossen', *Artilleristische Monatshefte*, Nr. 41, Berlin: 1910.

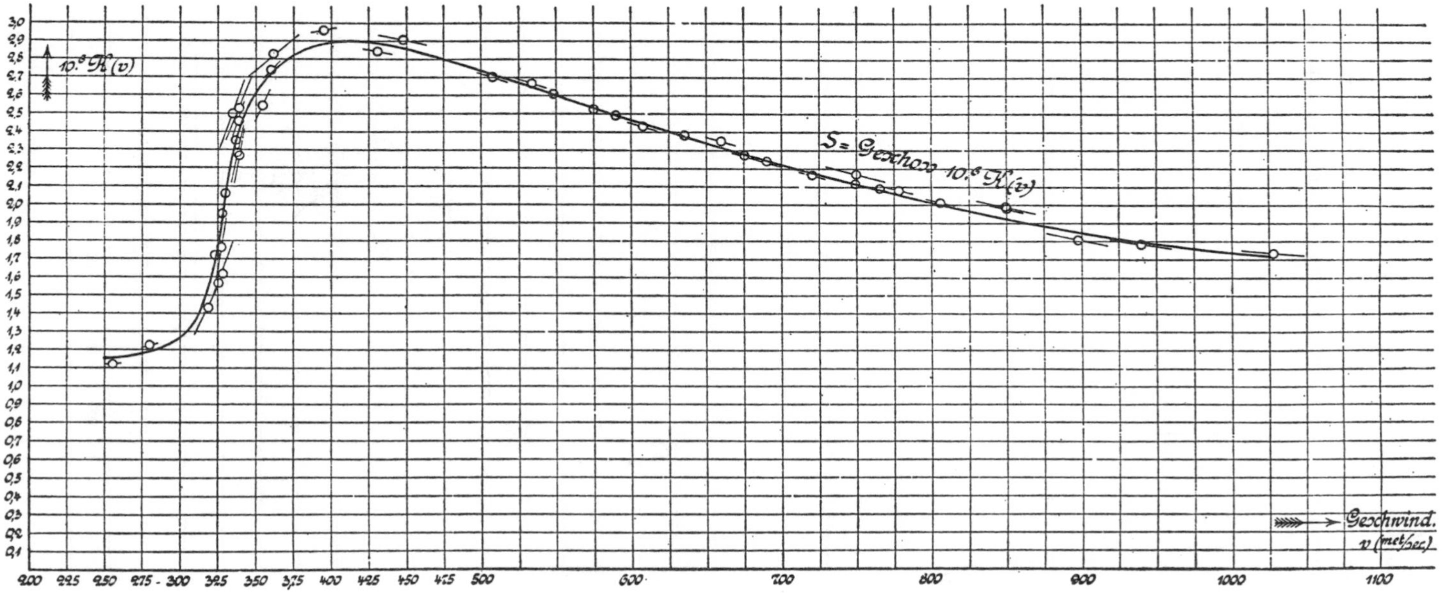

Fig. 122. Cranz's curve of transsonic and supersonic drag

Cranz leads one to the conviction that he tried to explain the sharp rise of the drag coefficient in the transsonic region. The idea of the shock tube was developed by him.†

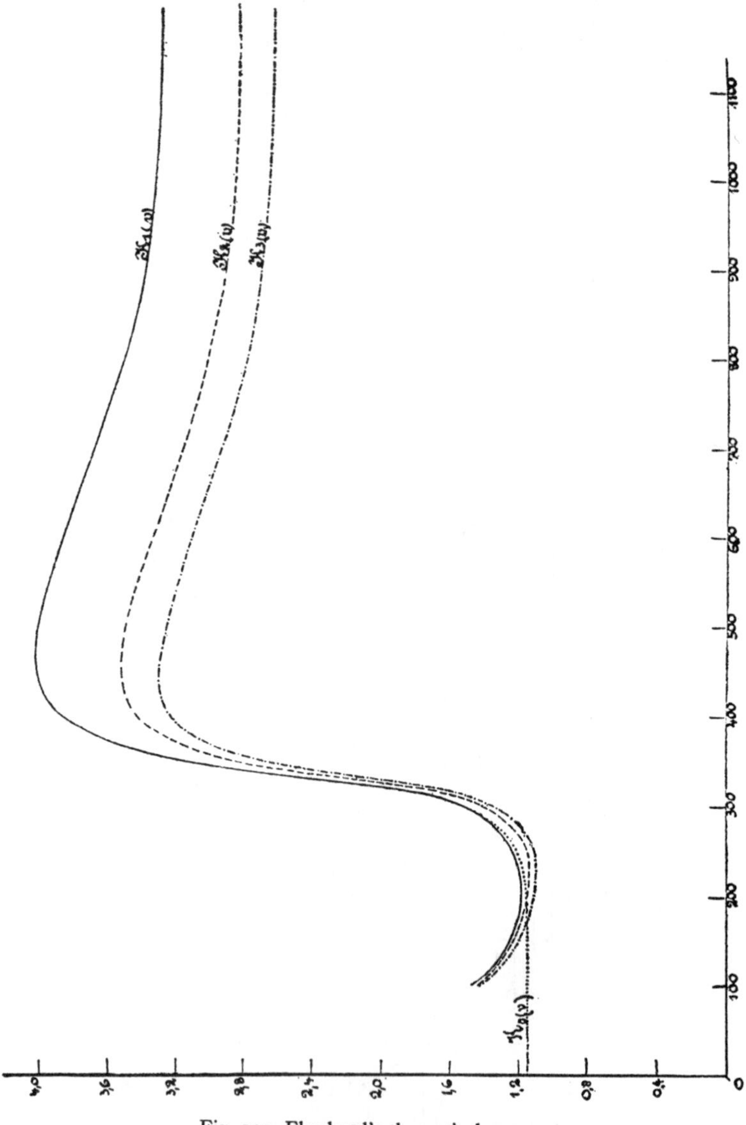

Fig. 123. Eberhard's theoretical curves

† ZS. f. Physik, 56, 170. C. Cranz und H. Scheridan: 1929.

E. Vallier's *Balistique expérimentale* (Paris, 1894) and F. Bashorth's *Mathematical treatise on the motion of projectiles* (1873, London), C. Cranz's *Äussere Ballistik* (1910, Leipzig) and other major publications on the subject tried to resolve the same problem of the Drag Wall, but none of them reached Siacci's heights. There was, however, one other researcher whose work deserves serious attention: O. v. Eberhard, who, in his 'Neue Versuche über Luftwiderstand gegen fliegende Geschosse und ihre Verwertung durch die theoretisch äussere Ballistik' (*Artilleristischer Monatshefte*, No. 71, September, 1912), gives an excellent mathematical theory and about 40 $C_D(v)$ curves, of which about 30 are theoretical and the rest experimental; many of the curves give the transsonic-supersonic drag pattern with surprising accuracy (Figure 123).

So, man managed to penetrate the secrets of the behaviour of the high-speed air particles, at least to a degree, at least for bullet-shaped configurations. And once he had established the existence of the drag 'wall', an achievement for which the credit and glory belong to ballisticians rather than to aerodynamicists, the problem of drag of an aircraft could be dealt with in a more refined way.

Transsonic compressibility effects on lift

We have so far been talking about the 'drag wall' of axi-symmetric bodies, at zero angle of attack. But most of the parts of an aircraft exposed in flight to air flow, and the aircraft as a whole, are non-symmetric. The prime aerodynamic function of the aircraft configuration is to create and maintain a sustentation (lifting) force; and the latter cannot be separated from the aerodynamic drag at any Mach number. The question is, what happens to the lift when the drag wall occurs?

Let us answer this question step-by-step. The following diagrams (Figure 124) show the interdependence of the shock-wave formation and pressure distribution of an aerofoil, at various Mach numbers and constant angle of attack.† At $M_0 = 0{\cdot}4$, the local velocity does not reach the value of speed of sound, therefore the aerofoil is free of shock waves.

At $M_0 = 0{\cdot}6$, near the point A of the upper surface, the local flow velocity reaches the value of the velocity of sound, and pressure is said to have become critical. After that, within a fairly short distance AB, the flow is supersonic and the pressure falls. At B, the supersonic flow

† N. S. Arzhanikov and V. N. Mal'tsev, *Aerodinamika* Moscow, 1956 (in Russian).

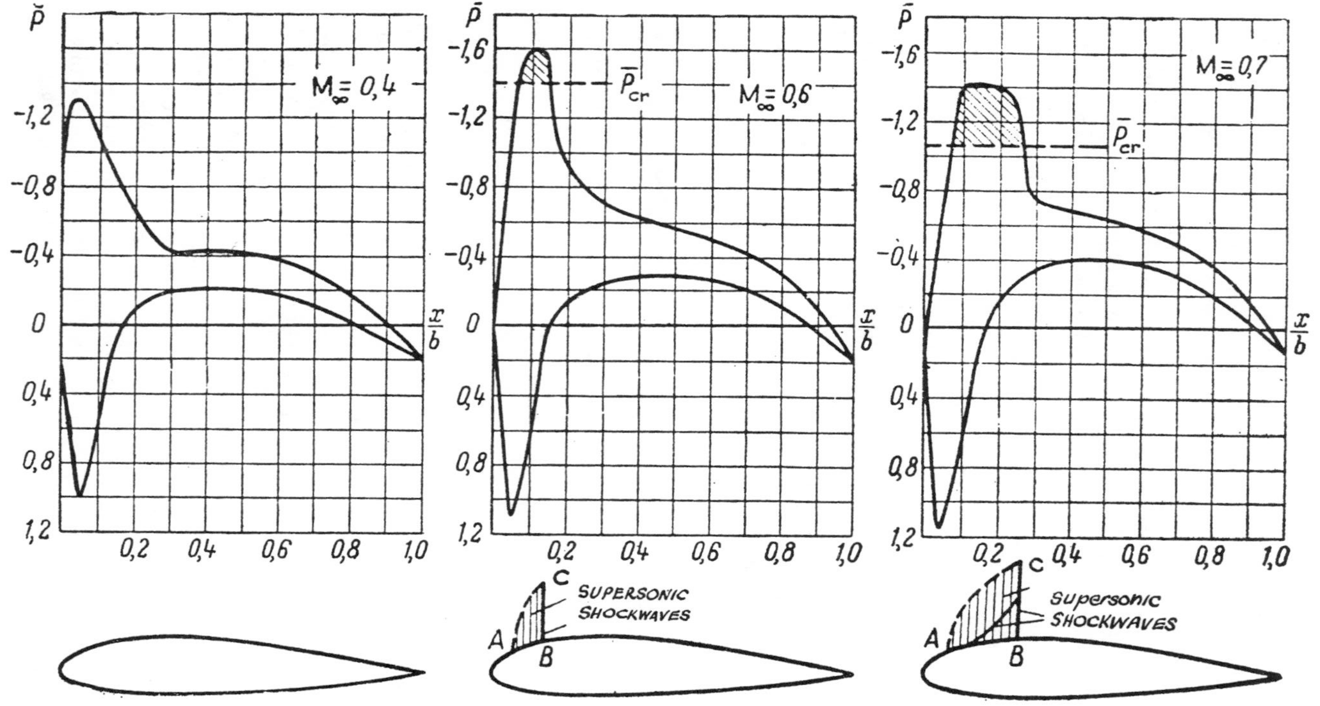

$\bar{p}$
$M_\infty = 0,4$
-1,2
-0,8
-0,4
0
0,4
0,8
1,2
0,2
0,4
0,6
0,8
1,0
$\frac{x}{b}$
$\bar{p}$
-1,6
$\bar{p}_{cr}$
$M_\infty = 0,6$
-1,2
-0,8
-0,4
0
0,4
0,8
1,2
0,2
0,4
0,6
0,8
1,0
$\frac{x}{b}$
C
SUPERSONIC
SHOCKWAVES
A
B
$\bar{p}$
-1,6
$M_\infty = 0,7$
-1,2
$\bar{p}_{cr}$
-0,8
-0,4
0
0,4
0,8
1,2
0,2
0,4
0,6
0,8
1,0
$\frac{x}{b}$
C
Supersonic
SHOCKWAVES
A
B

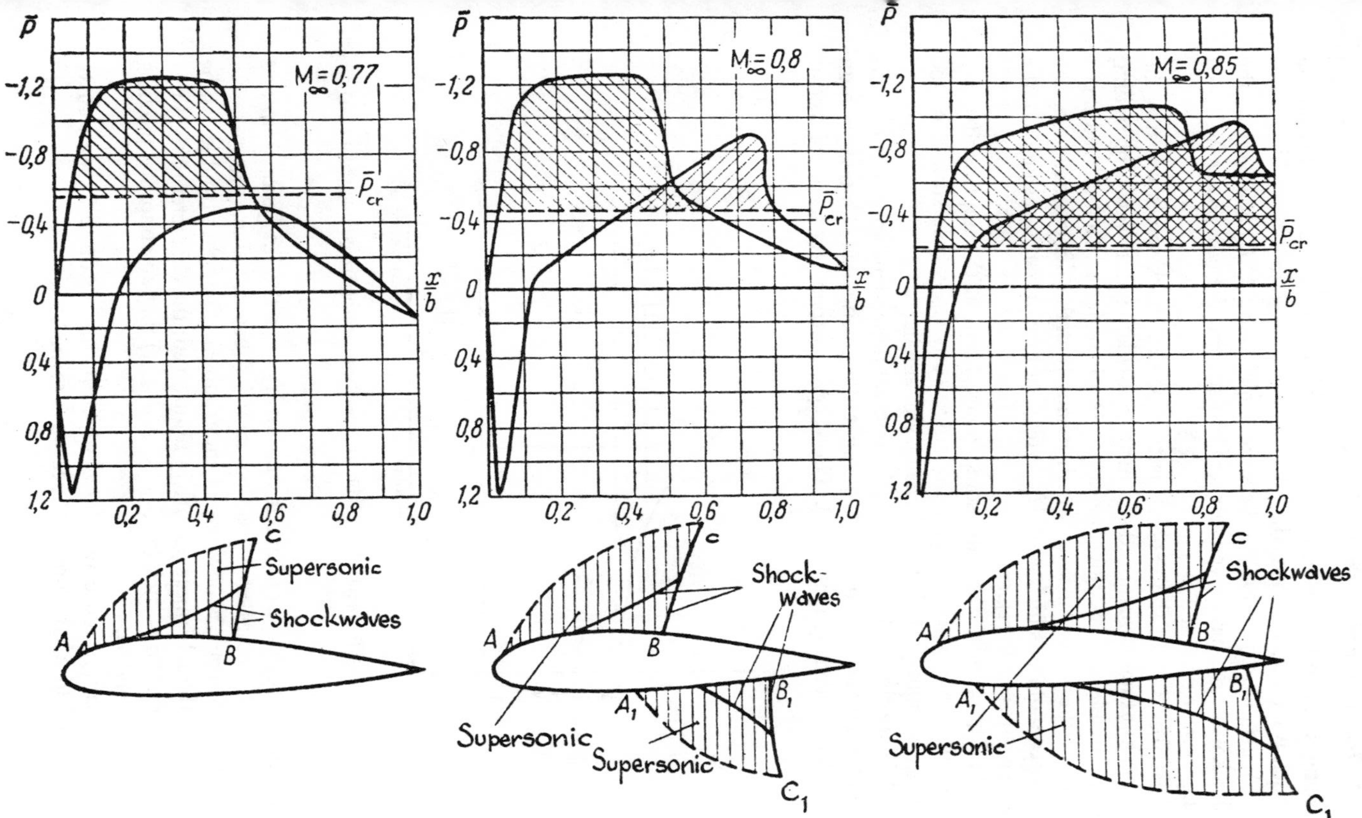

Fig. 124. Patterns of pressure distribution and shock waves on a wing

changes abruptly into subsonic, and pressure increases also abruptly: a shock wave (BC) is formed.

At $M_0 = 0{\cdot}7$ and $0{\cdot}77$, the supersonic zone (ABC) is enlarged; the shock wave BC has moved further downstream. We now have a λ-shaped shock-wave system.

At $M_0 = 0{\cdot}8$, the general pattern of the shock-waves remains the same, but they have moved still further downstream, towards the trailing edge. In addition, we now have a supersonic region and a λ-shaped shock also on the lower surface; which means that the lift coefficient of the wing cannot be as high as at $M_0 = 0{\cdot}7$, because pressure has fallen in the lower supersonic region.

At $M_0 = 0{\cdot}85$, the supersonic zones, both on the upper and lower surfaces, have grown larger, covering nearly the whole surface of the wing; obviously, its lift has become still smaller.

We are thus able to picture the general dependence of the lift coefficient of an aerofoil on the M_0 of flight at a given angle of attack, at high subsonic speeds (Figure 125). The very important fact to be noticed is

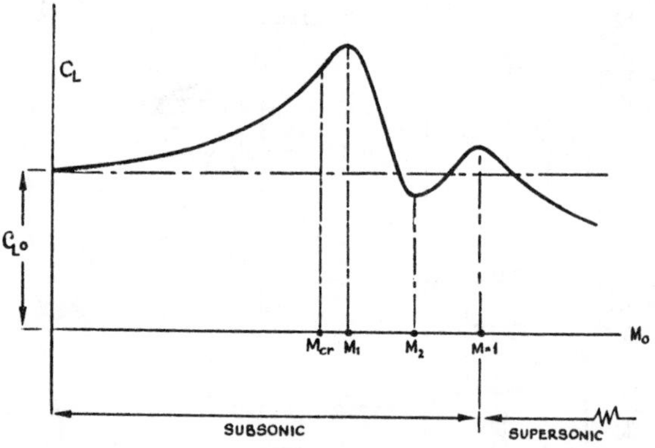

Fig. 125. The behaviour of the lift coefficient in the transsonic region

that in the region from about M_{cr} to $M_0 = 1$ the lift 'jumps' up and down, and becomes even less than its low subsonic value, C_{L0}.

That, however, is only one of the troubles. The other trouble can be seen from the schematic picture in Figure 126: the character of the pressure distribution is such (at $M_{cr} < M_0 < 1$) that an abnormal pitching moment develops.

And, finally, Figure 127 shows which points on the drag wall curve correspond to what kinds of shock wave formation.

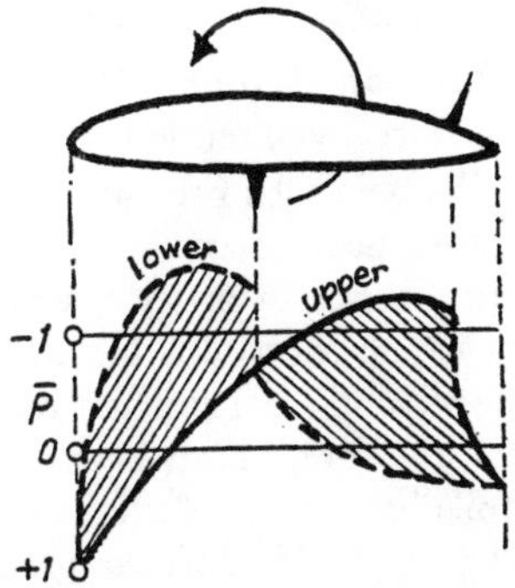

Fig. 126. The pressure distribution and pitching moment of an aerofoil at $M_{cr} < M_\infty < 1$

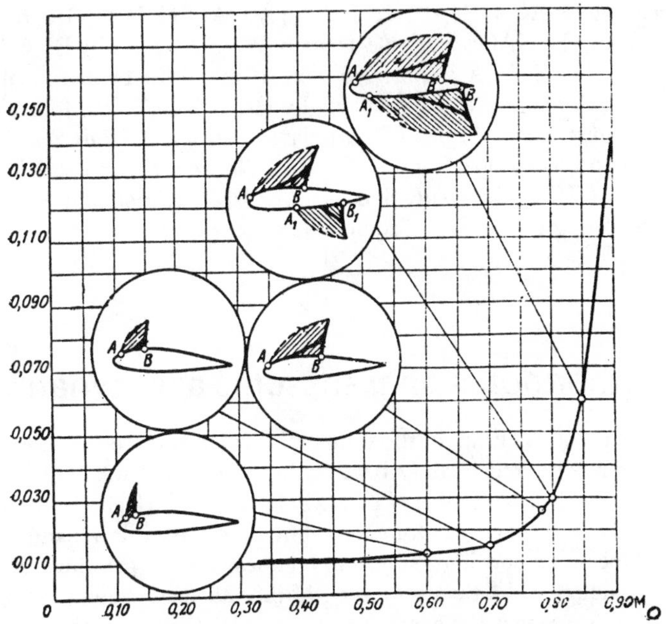

Fig. 127. A diagram showing the patterns of shock waves at various points on the 'drag wall' curve

The theoretical (mathematical) foundations of these experimental facts were under study a long time before the experimental evidence was accumulated. S. A. Tchaplygin was, perhaps, the first to publish (1904) a fundamental work in the field.† Then, in 1927, H. Glauert of

† The work in question is available in English: NACA TM No. 1063: 1944.

Cambridge published an important work called 'The Effect of Compressibility on the Lift of an Aerofoil' (R. and M., No. 1135) in which the relationship already mentioned emerged in the form $C_p = C_{p0} / \sqrt{1 - M_0^2}$, C_{p0} being the pressure coefficient at a speed low enough to consider the flow incompressible.

I must, then, refer to a series of articles, which steadily fade away from our memories, but which contributed in their time significantly to the formation of the theory and experimental methods of investigation of compressibility effects; they are: 'Tests in the Variable-Density Wind Tunnel of Related Airfoils having the Maximum Camber unusually far forward', by E. N. Jacobs and R. M. Pinkerton, NACA Report No. 537, 1935; 'Tests of Related Forward-Camber Airfoils in the Variable-Density Wind Tunnel', by E. N. Jacobs, R. M. Pinkerton and H. Greenberg, NACA Rep. No. 610, 1937; 'Airfoil Data Section Data obtained in the NACA Variable-Density Wind Tunnel', by E. N. Jacobs and I. Abbot, NACA Report No. 669, 1939; 'Two-Dimentional Subsonic Flow of Compressible Fluids', by Hsue-Shen Tsien, Journ. Aero. Sci., August, 1939; 'Tests of 16 Related Airfoils at High Speeds', by J. Stack and A. E. Doenhoff, NACA Rep. No. 492, 1934; 'Gas Flow past a Body at High Subsonic Speeds', by S. A. Khristianovich, CAGI Works, No. 481, 1940 – and some others.

Further notes on transsonic aerodynamics

The nature of gas flows in the transsonic region was still far from being adequately explored, mainly because by 'transsonic' was and is meant not just one Mach number, but a whole range of Mach numbers, say, from 0·82 to 1·2, because transsonic flows are mixed flows, and include both supersonic and high subsonic regimes. Figure 128 illustrates this clearly enough. The theory of wing of infinite span shows, and experiments endorse, that the sharp increase of C_D at certain subsonic Mach numbers can be delayed by means of sweeping the wing back by an angle χ; but it continues increasing up to about $M \geqslant 1$. . . .

A well designed aircraft wing should be capable of level flight at, say, M 0·82 or 0·85 without encountering serious compressibility flow collapse. But level flight is only one of the many conditions of work of a wing. For example, if high-speed manoeuvres are performed, an increase in the angle of attack is inevitable, therefore compressibility troubles must be anticipated at much lower free-stream Mach numbers.

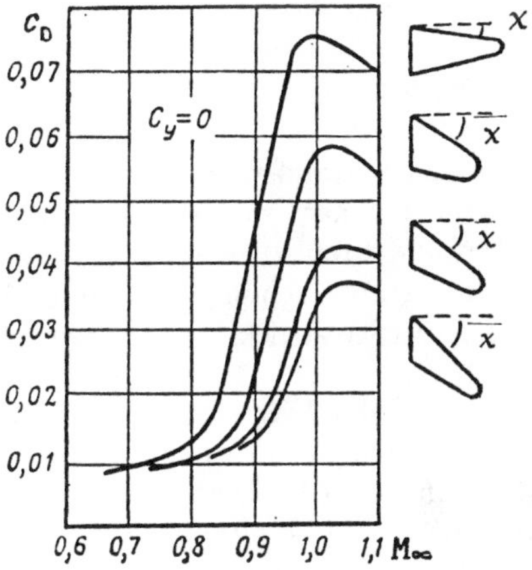

Fig. 128. Dependence of the 'drag wall' on the sweptback angle

The complex problems of transsonic flows were under theoretical and experimental investigations since about the second half of the thirties. More or less systematic, papers and books on the subject, began emerging, however, only after the Second World War.†

One of the most revealing works on the subject was published by Hans Wolfgang Liepmann in the *Journal of the Aeronautical Sciences* (December, 1946); it showed that the shock-pattern and the pressure distribution in a transsonic flow were strongly dependent upon the state of the boundary layer, and that a change from laminar to turbulent boundary layer at a given Mach number changed the flow pattern rather drastically.

These and many other phenomena required extensive experimental investigations, which could be carried out only in *transsonic wind-tunnels*. But the design and especially the calibration of such tunnels represented a very difficult task. I know from personal experience that even as recently as the period 1947–9 many people still doubted the possibility of transsonic wind tunnels at all. The reason was the unsteady nature of transsonic flows, and the 'wall' accompanied by the so-called 'blockage' and 'choking'. It can truly be said that the wind-tunnel designers had never faced so many headaches before. A completely new design philosophy was needed. It meant much time-consuming and costly efforts in Germany, France, Italy, USSR, USA

† See, for example, 'The hodograph method in trans-sonic flow'. M. J. Lighthill, Proc. Roy. Soc., London: 1947. *Studies on two-dimensional transonic flows of compressible fluid*. S. Tomotika and K. Tamada, University of Kyoto, Japan: 1946. *Transonic Wind Tunnel Testing*. B. H. Goethert, Pergamon Press, London: 1961.

and, no doubt, in other countries. But the rewards were far from satisfactory. Then, suddenly, someone, somewhere, remembered a long-forgotten article by K. Kondo, a Japanese,† which seemed to have the answer: the idea of slotting the walls of wind-tunnels. In due course, it led to the present transsonic wind-tunnels with perforated walls.

As usual, there were very many subsequent contributors to the theory of slotted and/or perforated walls. I will risk stating that the Deutsche Versuchsanstalt für Luftfahrtforschung, DVL, in Berlin-Aldershof, was the forerunner in this particular field. At any rate, this is my personal impression.

The USSR was miles behind Germany, but two years later – a very short time, indeed – I could see for myself that the Central'nyi Aero-Gidrodynamitcheskii Institut (the Central Aero- and Hydrodynamical Institute, near Moscow) had completed the design, construction and commission of one of the largest and most up-to-date transsonic wind-tunnels in the world at that time (Plate 21). It was created under the supervision of S. A. Khristianovich, S. A. Aristrakhov, B. V. Belyanin and V. G. Galperin. Sure enough, investigations in this tunnel led CAGI to the establishment of new properties of perforated walls, such as elimination of strong disturbances, irregularities in supersonic flows, stabilization of local Mach numbers, etc.‡

When the drag 'wall' occurs (at M $\sim$ 0·7–0·95), the blockage and choking became unacceptably high, and we simply do not know what we measure. If we now try to resolve the unpleasant problem of choking by means of keeping the flow open (open-jet tunnels), the unsteady and unsettled nature of near-sonic and transsonic flows causes more or less violent pulsations, the measuring instruments oscillate irregularly, and, again, we are faced with too many unknowns; besides, such tunnels are uneconomical in terms of driving power.

But man's knowledge has no sorrow that cannot be healed by man's thought; and a good thought once awakened does not slumber. Somebody somewhere (perhaps in Berlin, perhaps in Moscow, probably both in Berlin and Moscow) advanced the idea that the old enemy of aerodynamics – the boundary layer – may be the source of the above troubles; and that, therefore, the life-giving arrows of Eros should be aimed at this source. Hence the now well known ways of dealing with the problem: either to have moving walls, which would do away with the boundary

† 'Boundary Interference of Partly-closed Wind Tunnel'. K. Kondo, *Aeronautical Research Institute's Report* No. 137: 1936.

‡ G. L. Grozdovsky, A. A. Nikolsky, G. P. Svischev, G. I. Taganov: *Sverkhzvukovye tetcheniya gaza*, Moscow, 1967.

layer, or to have a large number of small holes through which the boundary layer could be sucked away (a series of 'boundary layer accelerators', jet streams by tiny pipes blowing along the wall strictly in the direction of the main flow).

In 1945, we learned, however, that during the war the German DVL had worked along different lines.† The Germans knew, as everyone else did working in the field, that at near-sonic and transsonic speeds the walls, model blockage, and throat choking interact and 'interspoil' almost everything; that the erratic shock-wave formation will prevent one from ever solving the problem of the transsonic wind-tunnel satisfactorily, unless something like longitudinal slots are incorporated.

The slots give the pressure pulsations a way of 'breathing out', and the flow an opportunity to calm down. But experience shows that the improvement is still greater when the walls are not slotted but perforated.

Fully perforated walls (see Plate 24), although not free from shortcomings, have now been accepted all over the world, because they keep the flow distortions, in the subsonic range, at an acceptably low level, thereby providing, among the other advantages, a more or less smooth transition from $M_0 < M_{cr}$ to $M_0 > M_{cr}$. The experience of the Department of Aeronautics of the City University, London, accumulated by M. M. Freestone and D. M. Sykes, and of other institutions, shows that such walls do have all these advantages. Plate 23 show that choking *is* almost completely eliminated: the unpleasant shock wave configurations and their locations are observable.

We can now say that the essential problems of experimental transsonic aerodynamics have been resolved. True enough, the difficulties are still formidable, but we at least know in what general direction we must go to deal with them. We know that this is a field where absolute technological perfection is an impossibility; but what has already been achieved shows that man is capable of solving most complex problems.

The question is often asked: do we really need to know the transsonic flow phenomena? Yes, we do. It is not accidental that pilots dislike transsonic flights so profoundly. When an aircraft goes through this region, unexpected changes can occur in the trim. Because of the appearance of a shock wave (or waves) on the upper surface of the wing, the point of action of the resultant lift may be suddenly displaced, thereby disturbing the relative locations of the lift and weight forces.‡

† See, for instance, *Transonic Wind Tunnel Testing*. B. H. Goethert, Pergamon Press, London: 1961.

‡ *Aerodynamics, selected topics in the light of their historical development*. Theodore von Karman, Cornell University Press, Ithaca, New York: 1957.

Even more unpleasant, if not dangerous, are the disturbances which can and do occur during high speed flight. Sometimes the pilot finds that his elevator or rudder is utterly ineffective. He moves the stick or the rudder pedals, but the aircraft fails to respond. This can be explained by the shock stalling of the fixed horizontal or vertical surfaces, in the presence of which the control surface moves in the wake and has no effect. At another time the pilot may find that the control surface is 'frozen'; apparently, the aerodynamic hinge moment has become so large that he is unable to overpower it. No complete explanation is known for this phenomenon; perhaps it has to do with the location of the shock wave. Some pilots say that they observe a shift of the control surfaces at a certain Mach number on a given airplane: the rudder, elevator, or aileron may suddenly leave its neutral position and jump to a deflected position without any action of the pilot.

Vibrations of the tail, or even of the whole aircraft, are often observed. Presumably in the mixed subsonic-supersonic flow over the wing the positions of the shock waves are not well defined; they may move back and forth. It has also been observed that, when the shock waves are produced on both the upper and lower surfaces of the wing or tail, they may move in opposite phase, which apparently makes the wake oscillate, and this oscillation is transferred to the wing or tail.

Further notes on supersonic fluidmechanics: superfluidity

I shall concentrate now mainly on supersonic shock-waves. Let us recall that the whole region outside the Mach cone (Figure 114d, p. 194) is free of disturbances. But what goes on in the immediate vicinity of the source of disturbance, where the latter can no longer be considered small?

To overcome the 'inconceivable stagnation point', d'Alembert proposed that bodies moving in fluids should have extremely sharp leading edges. But, first, you cannot have boats, aircraft fuselages and rockets with needle noses, aircraft wings and tailplanes with razor-blade-like leading edges, because of the possibility of fatal or serious accidents; second, even such noses and edges would not free aeronautics and space technology of shock waves.

Let us re-examine Figure 115, p. 196 and see what happens to Figure

129 at supersonic speeds. When millions and millions and millions of gas particles arrive in the region *O*, the pressure immediately before the leading edge (perhaps one could say, before the leading point) increases very sharply, the whole shaded region between the shock-wave and the edge becomes a region of compressed gas. Which means that oncoming

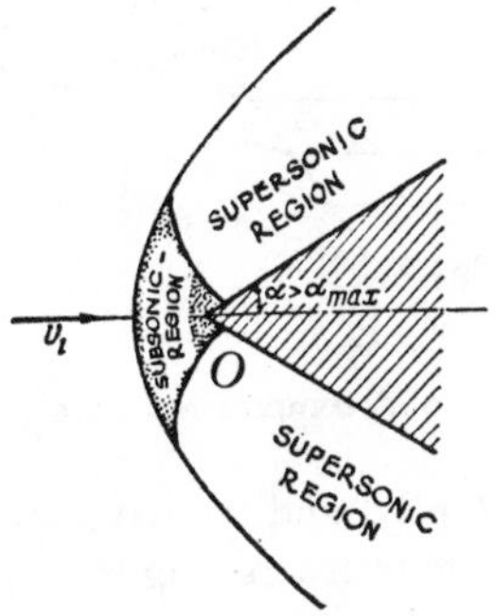

Fig. 129. Supersonic flow past a wedge (exaggerated)

particles cannot penetrate it with their supersonic velocities. In other words, the shock-wave divides the flow into two different regions: supersonic to the left, subsonic to the right. So, when particles cross the shock-wave, their velocities change from supersonic to subsonic. Beyond the wave, their further stagnation takes place in conditions close to adiabatic.

Rayleigh was the first to point out that this successive compression (first in the shock-wave, then in the subsonic region) results in a significantly lower pressure than in the base of a fully isentropic (adiabatic) compression from supersonic speeds to full stagnation.† It can also be shown that when gas particles cross the shock-wave, their entropy increases, and it follows from the energy equation that an increase in the entropy means a certain decrease in pressure. Thus, the existence of the shock-wave before a body of the type shown in Figure 115, page 196, leads to a certain decrease of the drag of the body.

In the case of a wedge the phenomena remain basically the same, but the size of the subsonic region is very much smaller, the stand-off distance of the wave becomes smaller and smaller as the free-stream Mach number increases; at a certain Mach number the wave is attached to the leading edge (if the latter is rounded, the wave remains detached, however small the stand-off distance). The other interesting fact is that as the free-stream Mach number increases, the shock-waves become less and less curved, while their angles of inclination become smaller and smaller.

† Proc. Roy. Soc., 84, 247: 1870.

Consider now a supersonic flow along a solid wall shown in Figure 130. When particles arrive at B, the wall BC compels them to change the

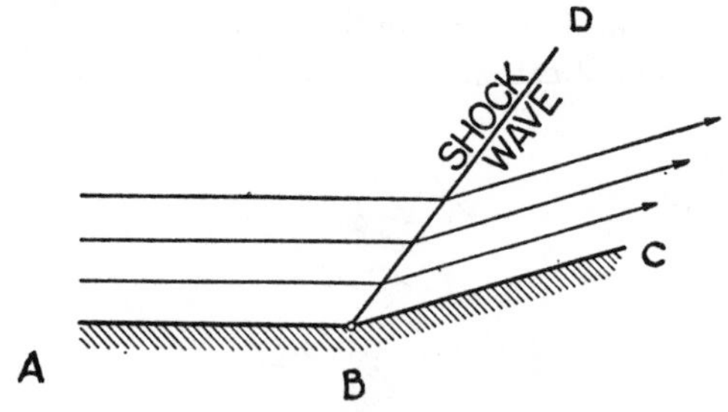

Fig. 130. A compression wave

direction suddenly and abruptly, and we have the compression shock-wave BD. If the wall has several corners (Figure 131), the shock becomes

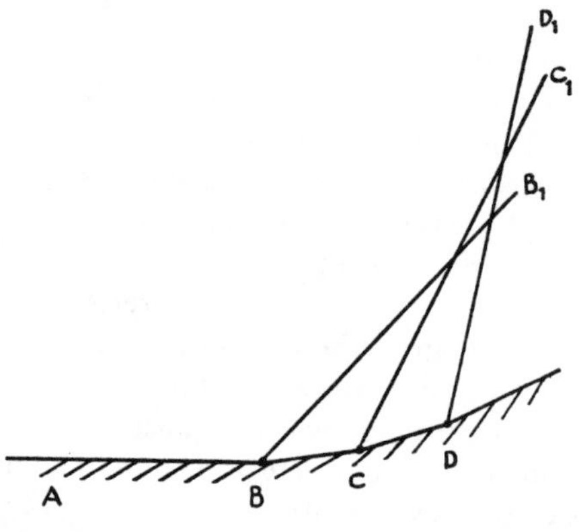

Fig. 131. Compression shock waves caused by successive corners

curved. One must emphasize, however, that the lines BB_1, CC_1, DD_1, etc., may not be strong enough to be considered as shock-waves: again, much depends on the free-stream Mach number, and the angle of the corner itself. This leads us back to the Mach lines we have already discussed.

An important and interesting feature of Mach waves is that when they are waves of expansion (rarefaction) they can never come together and reinforce each other, but they always spread apart like a fan; on the other hand, when they are waves of compression, they can and frequently do meet and reinforce each other; in such regions of reinforcement we have a large number of weak compressions adding up to form a front across which a definite and almost discontinuous compression takes place – a shock wave.†

If gas particles (Figure 132), arriving at *A* continued their motion

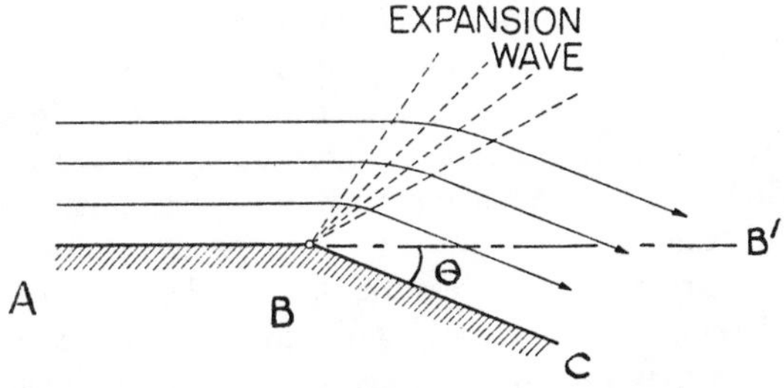

Fig. 132. Formation of expansion shock waves

along the straight line *BB*, the whole angular region θ would be in a state of vacuum, which is impossible physically and mathematically: fluids do not tolerate material discontinuities. What happens is this: for reasons opposite to those creating compression waves in Figures 115 and 130, there exists in Figure 132 a sudden drop in pressure at the point *B*. These changes occur gradually through a fan-like region known as the Prandtl-Meyer Region (or Expansion). It is interesting, theoretically, that there is no loss of energy in this expansion. On the other hand, expansion waves occur only in supersonic flows and are always oblique. (Plate 26 shows a fan of actual expansion waves on the wall of a wind tunnel.)

The first and fullest theory of compression and expansion waves was developed by Th. Meyer (see reference on page 207). From amongst the subsequent contributors to the theory, I wish to single out J. Ackeret, L. Prandtl and A. Busemann whose works always served and continue serving as the foundation, the walls and the roof of the edifice of modern supersonic gasdynamics. I cannot fail to add that, in my personal opinion and experience, among the more recent contributions to fluid-mechanics generally, and to gasdynamics in particular, by far the most outstanding in essence and beautiful in mathematical forms was the great book *Mekhanika Sploshnykh Sred* by Landau and Lifshitz, Moscow, 1954; it is difficult to imagine a professional pleasure superior to that experienced while reading this book.

It was, by the way, the 1944 issue of this book which gave modern fluidmechanics one of its most shining and newest chapters, the theory

† 'Gremlins and Barriers', an inaugural lecture delivered by Prof. A. D. Young on May 12, 1955, at Queen Mary College, University of London.

of superfluidity of Helium (which has nothing to do with supersonic gasdynamics, but since we are talking about Landau, I thought I might be allowed to say a few words about it).

There are in nature no ideal fluids. There is, however, one recently discovered 'gaseous liquid' which behaves much like an ideal fluid: Helium Two, He-2. Generally, helium is the most difficult of all the gases to liquefy; but it was liquefied in 1908 by K. H. Onnes of the University of Leiden, Holland. In 1938, a Soviet physicist, P. L. Kapitza, discovered the so-called 'superfluidity' of He-2, and in the 1941–4 period L. D. Landau (another Soviet scientist) developed its hydrodynamic theory.

Under one atmospheric pressure, liquid Helium freezes at $T = 14{\cdot}04°K$ and remains in this state down to the absolute zero. At $T = 2{\cdot}19°K$, however, it undergoes a certain change and has a point of transformation (λ-point) from one internal structure to another (Figure 133). Accordingly, liquid Helium at $T < 2{\cdot}19°K$ is called He-2. Liquid

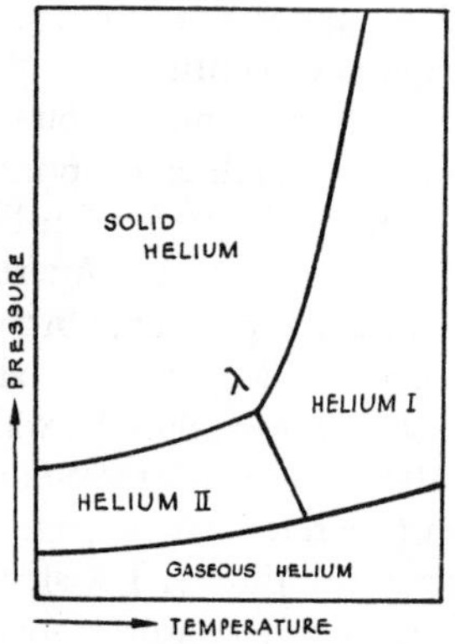

Fig. 133. Diagram of state of Helium

He behaves like liquid hydrogen, neon, air, etc., and may be considered as a perfectly normal fluid, its viscosity being of the order of $\mu \sim 10^{-5}$ poises.

The behaviour of He-2 is extraordinary: its heat conductivity exceeds that of copper and silver, but is not proportional to the temperature gradient; it can flow through thin capillaries, chinks, cracks and, indeed, through any holes without manifesting the slightest trace of viscosity (the latter is estimated to be $\mu \sim 10^{-11}$ poises, and cannot be measured by known methods); its flows do not transport heat, and are always potential. In short, He-2 behaves as an ideal fluid.

Landau's theory of superfluidity is based on the hypothesis that He-2 is a mixture of two different fluids: one of them possesses the property of superfluidity and the other behaves as a normal viscous fluid, so that the first flows through the second. The mathematical forms of the theory have the usual fluidmechanic pattern, but contain special terms.

Hypersonic gasdynamics

We now proceed to the branch of fluidmechanics called hypersonic gasdynamics, i.e. gasdynamics at $M \gtrsim 5$. The main peculiarity of this branch is the intensity of the shock-waves, which, by the way, have very much smaller Mach angles – are much more sweptback (see Plate 25: reproduced by courtesy of Hilton, W., *High Speed Aerodynamics*, Longmans Green, London, 1952).

In a sense, and in a degree, H. S. Tsien (now an important scientist in the People's Republic of China) can be called the father of hypersonic gasdynamics, because his theory provoked an almost immediate response† from several other 'fathers' and thereby paved the way to broader investigations. Hypersonic gasdynamics is closely associated with rockets and space vehicles, which have to fly in different physical conditions, from the *normal atmospheric continuum* to extlemely rarefied fringe layers of the atmosphere.

At very high altitudes, the atmosphere becomes so rarefied that it no longer behaves like a continuous fluid. The basic molecular character of air then gives rise to important modifications in aerodynamic and heat transfer phenomena. Continuum gas dynamics first had to be modified, then abandoned. The basic phenomena and theoretical approaches for highly rarefied flows are significantly different from those for flows that are only moderately rarefied. It is desirable therefore to divide rarefied gas dynamics into various flow regimes.‡

The characteristic flow in a highly rarefied gas is called 'free molecular flow'. In this regime the mean free path is large compared to the characteristic dimension of an aerodynamic body in the flow; and molecules that impinge on the body, and are then re-emitted from it will, in general, be far away from the body before they strike another molecule. It follows that the gas flow incident on the body is essentially undisturbed by the presence of the body. Aerodynamic and heat transfer characteristics

† 'Similarity laws of hypersonic flows'. H. S. Tsien, Journ. Math. Phys., 3, 25: 1946.
‡ NAVORD Report 1488, Vol 5, 1957.

depend only on the incident flow and the average momentum and energy interaction between incident molecules and the surface. Lift, drag, and heat transfer coefficients may be calculated in a straightforward way in terms of a few empirical surface interaction parameters. Experimental results, for the relatively small number of cases which have been investigated, are generally in good agreement with these theoretical predictions. However, these theoretical and experimental results have been confined to speeds and temperatures which are low enough to ensure that no internal molecular or atomic energy transformations of the gas molecules occur upon striking the surface. For very high velocity, high altitude applications it is to be expected that molecular dissociation, excitation, and even ionization will occur to some unknown extent on or perhaps near the surface.

The characteristic flow in a moderately rarefied gas is called 'slip flow'. This designation is taken from the phenomenon of 'slip', which is one of the important effects known to occur in a moderately rarefied gas flow and which is directly ascribable to its non-continuum, molecular structure. The gas layer immediately adjacent to a surface does not stick to it, but instead slips along it with a definite velocity proportional to the product of the wall shear stress and the molecular mean free path. A corresponding jump or discontinuity in the temperature is also known to exist.

The regime between slip and *free molecular flow* is called the 'transition flow regime'. It represents a density level for which the free path has the same general order of magnitude as the characteristic dimension of the flow.

The curves in Figure 134 show the physical regimes, or density levels, as functions of the Reynolds and Mach numbers and altitude.

The hypersonic gasdynamic theory of the first regime was developed by a number of authors: H. S. Tsien 'Superaerodynamics, Mechanics of Rarefied Gases', Journ. Aero. Sci., vol. 13, Dec., 1946; E. Sanger 'Gas Kinetik Sehr Grosser Flughohen', Schweizer Archiv für Angewandte Wissenschaft und Technik', vol. 16, 1950; M. Heineman 'Theory of Drag in Highly Rarefied Gases', Comm. on Pure and Applied Math., vol. 1, 1948; H. Ashley 'Applications of the Theory of Free Molecular Flow to Aeronautics', Journ. Aero. Sci., Feb., 1949, and others. As to the other regimes, I find it difficult to list their 'builders' in any logical order. Therefore, what follows below should be looked at as a more or less accidental order of dealing with the subject.

H. Julian Allen, one of the leaders of NASA's (National Aeronautics and Space Administration) Ames Research Centre near San Francisco, is the author of numerous papers on aerodynamics of atmosphere entry

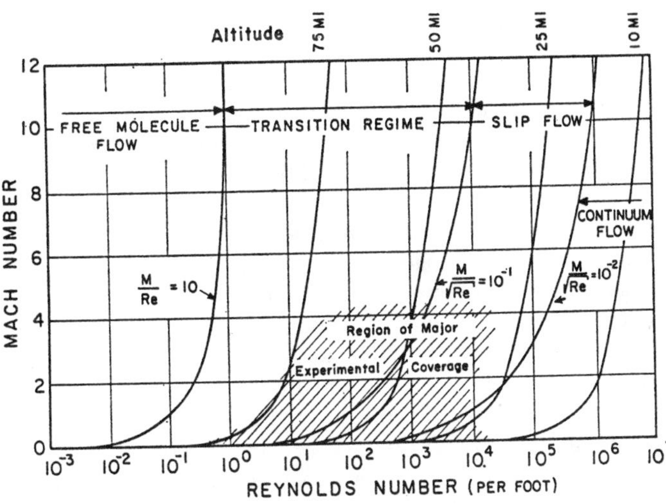

Fig. 134. Flow density regimes

vehicles and ballistic missiles. Back in 1958, he published a fundamental work on hypersonic flights and the problem of re-entry,† in which, among other ideas and concepts, he put forward the concept that blunted head configurations are needed to reduce heating during entry into the atmosphere of the earth. He warned, however, that such a configuration creates a strong detached shock-wave, which complicates the whole gas-dynamic theory. He also emphasized that rockets and space vehicles reach very high altitudes where the molecular mean free path may be of the same order as the dimensions of the body, therefore the study of the slip and free-molecular regimes was important.

He then looked at the problem from another angle. At hypersonic speeds, the stagnation pressure and temperature are so high that a degree of dissociation of the oxygen and nitrogen molecules of the air may occur; which, in turn, may lead to ionization. In these conditions, the accompanying convective heat exchange would be different from that in an 'ideal' gas. Moreover, the compressed and heated gas would become electrically conducting, and this would create practical difficulties for two-way radio communications.

Finally, Allen pointed out the necessity of studying not only the physical and chemical properties of the atmosphere, but also the nature of the high-energy particles, from meteors to cosmic rays.

These were, in effect, the basic philosophical concepts of Applied

† Journal of the Aeronautical Sciences, 25, No. 4: 1958.

Hypersonic Aerodynamics. As to his (Allen's) more specific contributions, I should like to mention his and A. J. Eggers' paper, 'A Study of the Motion and Aerodynamic Heating of Ballistic Missiles entering the Earth's Atmosphere at High Supersonic Speeds' (NACA Rep. 1381, 1958); the title of the paper shows the sort of problems they discuss – a very important contribution, indeed. Then, in November, 1962, at the NASA-University Conference on the Science and Technology of Space Exploration, Allen presented an excellent paper under the title 'Gas Dynamics Problems of Space Vehicles', dealing with the most acute problems of space flight and vehicle design. Even if this were his only contribution, the history and philosophy of space gasdynamics would remember him with deserved gratitude.

Alfred J. Eggers, another leading figure from the same NASA's Ames Research Centre in California, made very important contributions to the hypersonic aerodynamics of re-entry vehicles, of which I should like to single out his 'The Possibility of a Safe Landing'.† Rocket braking, atmospheric braking, impacting and grazing trajectories, hypersonic vehicle design considerations, motion and heating of variable-geometry ballistic vehicles in constant deceleration flight – these were the main headlines of the work. The reader should remember that there was at the time literally nothing like an established set of concepts on the subject: which made this particular contribution a world-wide event.

As I have said, the builders of hypersonic gasdynamics are listed here as they come to my mind, without any plan or preference. Lester Lees was, and remains, a prominent figure among them. Back in 1957, he outlined the main differences of the subject from ordinary gasdynamics.‡ Two years later, he published an outstanding work,§ in which he advanced a number of important propositions and methods of dealing with problems of convective heat transfer with mass addition and chemical reactions, of the use of aerodynamic lift during entry into the earth's atmosphere. The work included the results of experimental investigations of various head configurations at $M = 5{\cdot}8$. This, no doubt, was one of the finest works on the subject at the time, and put Lees' name among its creators.

Then there was Allace D. Hayes' and Ronald F. Probstein's series of articles – in due course integrated into a bulky book – on the whole range of hypersonic aerodynamics. With the appearance of their book (*Hyper-*

† 'Space Technology', Chapter 13, an excellent work of 54 pages, in the collective book *Space Technology*, John Wiley & Sons, Inc., New York: 1959. Also NACA TN No. 4046, 1957.

‡ 'Jet Propulsion', 27, No. 11: 1957.

§ 'Use of Aerodynamic Lift During Entry into the Earth's Atmosphere', ARS Journal, September, 1959.

sonic Flow Theory, Academic Press, New York, 1959), one could say that the formation of the new branch of fluidmechanics, hypersonic gasdynamics, was completed. But the list of its builders was, of course, much longer. It included Antonio Ferri, Alfred Busemann, George Sutton, M. D. Van Dyke, J. Stalder, N. Freeman, J. Fay and F. Riddell – and very many others in the USA, Canada, USSR, France, Japan, Great Britain and elsewhere.

The almost incredible development of military and space-exploration rocketry proved that the basic theories and experimental methods chosen were essentially correct. Let us have a closer look at some facts now known to the whole world.

Both in the American and Soviet programmes of exploration of the Moon, great importance was attached to the problem of re-entry into the atmosphere of the earth. It had to be remembered that the return speed of a lunar vehicle is $v = 36{,}000$ ft per second at an altitude of about 400,000 ft, and still higher just before it touches the upper fringes of the atmosphere. Several fundamental problems follow from here: (1) how to ensure that the vehicle remains within a narrow corridor between the airless perigees of the overshoot and undershoot boundaries, (2) what are the laws of aerodynamic resistance in the above discussed density levels, or regimes, (3) how to secure a degree of controllability of the ship during the crucial phase of re-entry, (4) what would be the stagnation region heating? etc., etc.

Space technology practice showed† that the theoretical and experimental solutions offered by the then still young hypersonic gasdynamics‡ were excellent in almost every respect. We now know, for instance, that the theory of the use of the lift-to-drag ratio was more than wise: it became a matter of success or failure. In the case of the Apollo, the centre of gravity of the vehicle was purposely offset from the longitudinal axis to provide aerodynamic lift and lift-to-drag ratio $L/D \sim 0{\cdot}3$. The orientation of the lift vector could be varied by using the reaction control system thrust to roll the vehicle about the drag vector, a stability axis roll. Thus, the lift vector could be oriented to any position between lift up and lift down to produce the necessary history of vertical and lateral aerodynamic force components needed for trajectory control.

The aerodynamic resistance of the atmosphere during re-entry causes (1) severe vehicle deceleration, (2) severe aerodynamic loads, and (3) severe aerodynamic heating. With reference to the latter, compression

† See, for example, 'Apollo Program', Space Division of North American Rockwell Corporation: 1968.

‡ Proceedings of the NASA-University Conference on the Science and Technology of Space Exploration, vol 2, Chicago: 1962.

between the main shock-wave and the body's leading surface is so intense that the maximum temperature in the stagnation region may be up to 5800°F (Apollo 13). And at such temperatures the air becomes a vibrating, dissociating and ionized gas. Because of the presence of electrons and ions, there are strong influences on the energy and mass transport even at a considerably low level of ionization. The termal conductivity increases rapidly – and so on, and so forth.

It should not be thought, however, that hypersonic gasdynamics is the science of rocketry and space technology alone. No, it has a much wider value. We shall have new types of space vehicles, and we shall in future have also hypersonic aircraft. Who even dreamed only (say) 25 to 30 years ago that there would be supersonic aircraft like the Concorde and Tu-144? But today they are with us. There is no doubt whatsoever that hypersonic air transport, too, is on its way. Therefore, hypersonic gasdynamics is more than merely a science of rocketry and re-entry vehicles.

The universal matter-energy continuity

Mathematically, by space is meant simply a three-dimensional region. But the space I am going to discuss now is more than that: it is 'filled' with matter and energy in numerous states and forms of manifestation. Moreover, *space* (in the mathematical sense of the word), *matter, energy and time do not exist separately and independently*, none of them is even imaginable without the others.

In this formulation, the concept 'empty space' loses its meaning, and the concept of space exploration becomes meaningful. For, indeed, the Latin word *materia* means that which exists everywhere in the universe. The Greek word *energeia* means, literally, one of the main properties of matter, of materia. Man, animals, the earth, planets, stars, galaxies, atmosphere, water, fire, light, darkness, etc., are different forms and states of existence of matter and energy. In this sense, the most fundamental property of space – of the universe – is its universal matter and energy continuity.

The history of science shows, wrote Louis de Broglie,† that progress in knowledge has constantly been hampered by the tyrannical influence of certain conceptions that finally came to be considered as dogmas; for this reason, it is proper to submit periodically to a very searching examination, principles that we have come to assume without discussion.

† Revolution in Physics: 1953.

And this was precisely what was applied to the World Aether concept.

The blow was delivered by Albert Einstein's field theory,† and by his attempt to create a unified field theory based upon the space-time four-dimensional geometry. This theory represents, first of all, a revision of Newton's theory of gravitation, in which forces act upon bodies, but nothing happens in the space between them; while in Einstein's general theory of relativity forces acting upon bodies are not so important as that which takes place in the space between them, the gravitational field (which propagates with finite speed). Einstein succeeded in formulating the law governing this field, and then derived from it the (Newtonian) laws of motion.

Every corner of the universe is occupied by matter and/or energy. But matter, energy and fields are inseparable. Wherever and whenever one of the three is present, the other two also are present. If we could destroy or create matter, we would thereby destroy or create energy and fields. If we could 'remove' the electromagnetic fields, nothing would bind atoms to molecules and electrons to atoms. If there were no gravitational fields, nothing would keep our earth in its orbit, stars in their galaxies, and the whole world would collapse.

Matter exists not only in the form of material bodies. Light and radio-waves have no weight or other tangible properties, therefore we cannot call them material bodies. But they have such properties as intensity, wavelengths, frequency, time-dependence, etc., which can be measured.

Thus, space and time are 'filled' and transversed with and by matter, by energy fields. Empty space, or absolute vacuum, does not exist, cannot exist, cannot even be imagined. Nor can we imagine an empty time.

The term light properly refers to the range of electromagnetic radiation frequencies associated with vision. The wavelengths of visible light extend approximately from 4000 ångstroms (extreme violet) to 7700 ångstroms (extreme red). But Max Planck showed that radiation of any kind is a flow of energy; the basic unit of energy emitted in a single wave motion of radiation is 'quantum', otherwise known as 'photon'.

So, light is a flow of photons. And the mass at rest of a photon is zero! But this does not mean that the photon is non-existent. Indeed, it does exist: in motion. That is, there are no photons at rest, there are only photons in motion. Which is the same as saying that to stop light from propagation would mean to destroy it, to destroy matter, energy, electromagnetic fields, which is an impossibility.

The other striking characteristic of photons, of light, is that they move at 300,000 km/sec. Furthermore, this must be the *maximum maximorum*

† Albert Einstein, Collected Works, six volumes (in Russian), published by 'Nauka', Moscow. Also 'Mein Weltbild', Frankfurt-am-Main: 1955.

speed in the world; no material body in motion can reach this speed. Why? Because otherwise, when it catches up with the photon, it would be at rest in relation to light, therefore it would cease to exist (which is an impossibility).

Anyone who has ever thrust a glass prism into a sunbeam and seen rainbow colours of the solar spectrum refracted on a screen has looked upon the whole range of visible light. For the human eye is sensitive only to the narrow band of radiation that falls between the red and the violet.

The wavelength of red light is 0·00007 cm, and that of violet light 0·00004 cm. But the sun emits also other kinds of radiation: infra-red rays of 0·00008 to 0·032 cm; ultra-violet rays of 0·00003 to 0·000001 cm; X-rays which are shorter than ultra-violet rays; and other kinds of rays. Let me repeat, the world and we are literally criss-crossed by them, without our being able to observe them. Furthermore, we just do not know what proportions of what class of radiation from outer space are prevented by the atmosphere from reaching the earth.

Max Planck showed that each quantum carries an amount of energy $E = h\nu$, where ν is the frequency of the radiation and h is Planck's Constant (= 0·000000000000000000000000006624), which means that energy exists everywhere in the universe; and Einstein showed that energy is equivalent to mass, which means that matter exists everywhere in the universe.

But every material particle attracts every other particle with a force that is directly proportional to the product of their masses and inversely proportional to the square of the distance between them. The implication of this is that every cubic millimeter of space is traversed not only by radiation waves, but also by waves of gravitation. We do not see them, but they are there.

Imagine a spaceship flying at a speed greater than that of light. It would first catch up with the light it emitted yesterday, then with the light emitted the day before yesterday, and so on. And the crew of the ship would be moving, in relation to time, from the present into the past, i.e. becoming younger and younger.

On the other hand, so long as speed remains constant, the energy of motion also remains constant. But when the speed increases, the energy also increases. The question is, where does the body get this additional energy? Does this mean that in order to move faster our spaceship must have a magic source of energy?

We know from elementary mechanics that the greater the mass of a rocket, the more difficult it is to accelerate it; vice versa, the smaller the mass, the easier it is to accelerate it. But up to what speed? The answer is as before: the space vehicle cannot reach the speed of light – unless, of course, its mass diminishes completely. But this is the one thing we cannot do with our spaceship: a non-existing vehicle! So the mass remains, and yet it approaches the speed of light – remember, this is an imaginary case. What does this mean? It means that the mass – or the weight – of the vehicle becomes 'very great', so great that no further effort can accelerate it still further, up to the speed of light, when it becomes 'infinitely massive'.

The vehicle gets more massive only when its speed – and, consequently, its energy of motion – increases. Therefore the 'additional mass' is 'brought in' by the energy of motion; an inflow of energy is equivalent to an inflow of mass. But how much is that 'additional mass'? The answer is given by Einstein's formula

$$\boxed{E = m.c.^2},$$

where E is energy, m is mass and c is the speed of light. This revolutionary formula was derived by Albert Einstein on three pages of his historic work 'Ist die Trägheit eines Körpers von seinem Energieinhalt abhängig?' (Annalen der Physik, Bd. 18, 639, 1905). It shows that the laws of classical mechanics need to be modified in order to come into line with the theory of relativity. To give just one example, the kinetic energy of a body $mv^2/2$ becomes $mc^2/\sqrt{1 - v^2/c^2}$.

I should like to call $v^2/c^2 = \epsilon$ 'Einstein's Number'. When $v \to c$, then $\epsilon \to 1$, the kinetic energy tends to become infinitely great, which is a physical and mechanical impossibility. We need not, however, worry because nothing in the world can move with the speed of light; in all cases of modern space flights $\epsilon = v^2/c^2 <<< 1$, so that the formula reduces to mc^2.

Imagine a spaceship, weighing half a ton, launched towards the moon at $v = 11{\cdot}2$ km/sec; its kinetic energy is immense, by ordinary standards; its mass increases, however, by only one milligram! Half a ton and one milligram: there are so far no instruments capable of recording such a change. But one day man will have photon rockets capable of imparting speeds of the order of, say, 260,000 km/sec, and then the mass of the vehicle will be doubled.

And why do we need such high speeds? The answer springs from the character of the universe itself. You may well imagine that our globe is a flat in the third storey of a house which we call the solar system. The house is located in an outlying part of a vast stellar community numbering some 150,000 million stars. And even the nearest star beyond the sun, Alpha Centauri, is so far away that 4·3 light years are needed to cover the distance, the equivalent of 270,000 times the distance between the earth and the sun.

To reach even this nearest star by a spaceship propelled by a rocket of the basic scheme we use today, hundreds of thousands of years would be needed to cover the distance and return. But, to put it mildly, man does not live that long. Therefore, the solution of the problem will be sought along new lines; and the idea of a photon rocket represents such a new line.

Let me now recall that an electron is an elementary particle with a negative charge within an atom. Then there are in the same atom the so-called protons; the mass of a proton is 1800 times greater than that of an electron, but has a positive charge.

Protons and electrons exist in all material bodies, in matter. But the latter contains also elementary particles called positrons and anti-protons. The mass of a positron is equal to that of an electron, but has a positive charge; the mass of an anti-proton is equal to the mass of a proton, but has a negative charge. Generally, an anti-particle has the same mass as its corresponding particle, and is identical in almost every respect, but has the opposite charge. When a particle meets its anti-particle, they annihilate each other completely, with the release of their energy in the form of photons, and this leads to the idea of a photon rocket.

Suppose we know how to create and to store matter and anti-matter in flight. And suppose we know how to accelerate them to a point of collision in front of a very large parabolic mirror. Then they will annihilate each other and release an immense amount of photons, which would be reflected by a mirror into a parallel photon stream, with the speed of light $c = 300{,}000$ km/sec. This stream would create a photon reactive force – photon thrust – and we would have a photon rocket.

The flight of a photon spaceship can be divided into three major stages: acceleration, steady inertial flight, and deceleration. If throughout the first stage the spaceship maintains the gravitational acceleration of the earth, it will in about 17 months reach the speed of 250,000 km/sec, or 83 per cent of the speed of light, and cover a distance of 7300 milliard kilometres, or 18 per cent of the distance to Alpha Centauri. After that the second stage would begin: the photon engine would be switched off and the flight would continue by inertia, at 250,000 km/sec, for 3 years 4 months. Finally, the deceleration systems would be switched on, and the third stage would commence; its duration in time and distance would be the same as for the first stage. So, the total duration of flight to the nearest star would be 6 years 2 months in one way, or 12 years 4 months in both ways.

Index

A CATALOG OF SELECTED

DOVER BOOKS

IN SCIENCE AND MATHEMATICS

A CATALOG OF SELECTED

DOVER BOOKS

IN SCIENCE AND MATHEMATICS

QUALITATIVE THEORY OF DIFFERENTIAL EQUATIONS, V.V. Nemytskii and V.V. Stepanov. Classic graduate-level text by two prominent Soviet mathematicians covers classical differential equations as well as topological dynamics and ergodic theory. Bibliographies. 523pp. 5⅜ x 8½. 65954-2 Pa. $14.95

MATRICES AND LINEAR ALGEBRA, Hans Schneider and George Phillip Barker. Basic textbook covers theory of matrices and its applications to systems of linear equations and related topics such as determinants, eigenvalues and differential equations. Numerous exercises. 432pp. 5⅜ x 8½. 66014-1 Pa. $10.95

QUANTUM THEORY, David Bohm. This advanced undergraduate-level text presents the quantum theory in terms of qualitative and imaginative concepts, followed by specific applications worked out in mathematical detail. Preface. Index. 655pp. 5⅜ x 8½. 65969-0 Pa. $14.95

ATOMIC PHYSICS (8th edition), Max Born. Nobel laureate's lucid treatment of kinetic theory of gases, elementary particles, nuclear atom, wave-corpuscles, atomic structure and spectral lines, much more. Over 40 appendices, bibliography. 495pp. 5⅜ x 8½. 65984-4 Pa. $13.95

ELECTRONIC STRUCTURE AND THE PROPERTIES OF SOLIDS: The Physics of the Chemical Bond, Walter A. Harrison. Innovative text offers basic understanding of the electronic structure of covalent and ionic solids, simple metals, transition metals and their compounds. Problems. 1980 edition. 582pp. 6⅛ x 9¼. 66021-4 Pa. $16.95

BOUNDARY VALUE PROBLEMS OF HEAT CONDUCTION, M. Necati Özisik. Systematic, comprehensive treatment of modern mathematical methods of solving problems in heat conduction and diffusion. Numerous examples and problems. Selected references. Appendices. 505pp. 5⅜ x 8½. 65990-9 Pa. $12.95

A SHORT HISTORY OF CHEMISTRY (3rd edition), J.R. Partington. Classic exposition explores origins of chemistry, alchemy, early medical chemistry, nature of atmosphere, theory of valency, laws and structure of atomic theory, much more. 428pp. 5⅜ x 8½. (Available in U.S. only) 65977-1 Pa. $11.95

A HISTORY OF ASTRONOMY, A. Pannekoek. Well-balanced, carefully reasoned study covers such topics as Ptolemaic theory, work of Copernicus, Kepler, Newton, Eddington's work on stars, much more. Illustrated. References. 521pp. 5⅜ x 8½. 65994-1 Pa. $12.95

PRINCIPLES OF METEOROLOGICAL ANALYSIS, Walter J. Saucier. Highly respected, abundantly illustrated classic reviews atmospheric variables, hydrostatics, static stability, various analyses (scalar, cross-section, isobaric, isentropic, more). For intermediate meteorology students. 454pp. 6⅛ x 9¼. 65979-8 Pa. $14.95

ASYMPTOTIC METHODS IN ANALYSIS, N.G. de Bruijn. An inexpensive, comprehensive guide to asymptotic methods–the pioneering work that teaches by explaining worked examples in detail. Index. 224pp. 5⅜ x 8½. 64221-6 Pa. $7.95

OPTICAL RESONANCE AND TWO-LEVEL ATOMS, L. Allen and J. H. Eberly. Clear, comprehensive introduction to basic principles behind all quantum optical resonance phenomena. 53 illustrations. Preface. Index. 256pp. 5⅜ x 8½. 65533-4 Pa. $8.95

COMPLEX VARIABLES, Francis J. Flanigan. Unusual approach, delaying complex algebra till harmonic functions have been analyzed from real variable viewpoint. Includes problems with answers. 364pp. 5⅜ x 8½. 61388-7 Pa. $9.95

ATOMIC SPECTRA AND ATOMIC STRUCTURE, Gerhard Herzberg. One of best introductions; especially for specialist in other fields. Treatment is physical rather than mathematical. 80 illustrations. 257pp. 5⅜ x 8½. 60115-3 Pa. $7.95

APPLIED COMPLEX VARIABLES, John W. Dettman. Step-by-step coverage of fundamentals of analytic function theory–plus lucid exposition of five important applications: Potential Theory; Ordinary Differential Equations; Fourier Transforms; Laplace Transforms; Asymptotic Expansions. 66 figures. Exercises at chapter ends. 512pp. 5⅜ x 8½. 64670-X Pa. $12.95

ULTRASONIC ABSORPTION: An Introduction to the Theory of Sound Absorption and Dispersion in Gases, Liquids and Solids, A.B. Bhatia. Standard reference in the field provides a clear, systematically organized introductory review of fundamental concepts for advanced graduate students, research workers. Numerous diagrams. Bibliography. 440pp. 5⅜ x 8½. 64917-2 Pa. $11.95

UNBOUNDED LINEAR OPERATORS: Theory and Applications, Seymour Goldberg. Classic presents systematic treatment of the theory of unbounded linear operators in normed linear spaces with applications to differential equations. Bibliography. 199pp. 5⅜ x 8½. 64830-3 Pa. $7.95

LIGHT SCATTERING BY SMALL PARTICLES, H.C. van de Hulst. Comprehensive treatment including full range of useful approximation methods for researchers in chemistry, meteorology and astronomy. 44 illustrations. 470pp. 5⅜ x 8½. 64228-3 Pa. $12.95

CONFORMAL MAPPING ON RIEMANN SURFACES, Harvey Cohn. Lucid, insightful book presents ideal coverage of subject. 334 exercises make book perfect for self-study. 55 figures. 352pp. 5⅜ x 8¼. 64025-6 Pa. $11.95

OPTICKS, Sir Isaac Newton. Newton's own experiments with spectroscopy, colors, lenses, reflection, refraction, etc., in language the layman can follow. Foreword by Albert Einstein. 532pp. 5⅜ x 8½. 60205-2 Pa. $12.95

GENERALIZED INTEGRAL TRANSFORMATIONS, A.H. Zemanian. Graduate-level study of recent generalizations of the Laplace, Mellin, Hankel, K. Weierstrass, convolution and other simple transformations. Bibliography. 320pp. 5⅜ x 8½. 65375-7 Pa. $8.95

THE ELECTROMAGNETIC FIELD, Albert Shadowitz. Comprehensive undergraduate text covers basics of electric and magnetic fields, builds up to electromagnetic theory. Also related topics, including relativity. Over 900 problems. 768pp. 5⅝ x 8¼. 65660-8 Pa. $18.95

FOURIER SERIES, Georgi P. Tolstov. Translated by Richard A. Silverman. A valuable addition to the literature on the subject, moving clearly from subject to subject and theorem to theorem. 107 problems, answers. 336pp. 5⅜ x 8½. 63317-9 Pa. $9.95

THEORY OF ELECTROMAGNETIC WAVE PROPAGATION, Charles Herach Papas. Graduate-level study discusses the Maxwell field equations, radiation from wire antennas, the Doppler effect and more. xiii + 244pp. 5⅜ x 8½. 65678-0 Pa. $6.95

DISTRIBUTION THEORY AND TRANSFORM ANALYSIS: An Introduction to Generalized Functions, with Applications, A.H. Zemanian. Provides basics of distribution theory, describes generalized Fourier and Laplace transformations. Numerous problems. 384pp. 5⅜ x 8½. 65479-6 Pa. $11.95

THE PHYSICS OF WAVES, William C. Elmore and Mark A. Heald. Unique overview of classical wave theory. Acoustics, optics, electromagnetic radiation, more. Ideal as classroom text or for self-study. Problems. 477pp. 5⅜ x 8½. 64926-1 Pa. $13.95

CALCULUS OF VARIATIONS WITH APPLICATIONS, George M. Ewing. Applications-oriented introduction to variational theory develops insight and promotes understanding of specialized books, research papers. Suitable for advanced undergraduate/graduate students as primary, supplementary text. 352pp. 5⅜ x 8½. 64856-7 Pa. $9.95

A TREATISE ON ELECTRICITY AND MAGNETISM, James Clerk Maxwell. Important foundation work of modern physics. Brings to final form Maxwell's theory of electromagnetism and rigorously derives his general equations of field theory. 1,084pp. 5⅜ x 8½. 60636-8, 60637-6 Pa., Two-vol. set $25.90

AN INTRODUCTION TO THE CALCULUS OF VARIATIONS, Charles Fox. Graduate-level text covers variations of an integral, isoperimetrical problems, least action, special relativity, approximations, more. References. 279pp. 5⅜ x 8½. 65499-0 Pa. $8.95

HYDRODYNAMIC AND HYDROMAGNETIC STABILITY, S. Chandrasekhar. Lucid examination of the Rayleigh-Benard problem; clear coverage of the theory of instabilities causing convection. 704pp. 5⅝ x 8¼. 64071-X Pa. $14.95

CALCULUS OF VARIATIONS, Robert Weinstock. Basic introduction covering isoperimetric problems, theory of elasticity, quantum mechanics, electrostatics, etc. Exercises throughout. 326pp. 5⅜ x 8½. 63069-2 Pa. $9.95

DYNAMICS OF FLUIDS IN POROUS MEDIA, Jacob Bear. For advanced students of ground water hydrology, soil mechanics and physics, drainage and irrigation engineering and more. 335 illustrations. Exercises, with answers. 784pp. 6⅛ x 9¼. 65675-6 Pa. $19.95

NUMERICAL METHODS FOR SCIENTISTS AND ENGINEERS, Richard Hamming. Classic text stresses frequency approach in coverage of algorithms, polynomial approximation, Fourier approximation, exponential approximation, other topics. Revised and enlarged 2nd edition. 721pp. 5⅜ x 8½. 65241-6 Pa. $15.95

THEORETICAL SOLID STATE PHYSICS, Vol. 1: Perfect Lattices in Equilibrium; Vol. II: Non-Equilibrium and Disorder, William Jones and Norman H. March. Monumental reference work covers fundamental theory of equilibrium properties of perfect crystalline solids, non-equilibrium properties, defects and disordered systems. Appendices. Problems. Preface. Diagrams. Index. Bibliography. Total of 1,301pp. 5⅜ x 8½. Two volumes. Vol. I: 65015-4 Pa. $16.95
Vol. II: 65016-2 Pa. $16.95

OPTIMIZATION THEORY WITH APPLICATIONS, Donald A. Pierre. Broad spectrum approach to important topic. Classical theory of minima and maxima, calculus of variations, simplex technique and linear programming, more. Many problems, examples. 640pp. 5⅜ x 8½. 65205-X Pa. $16.95

THE CONTINUUM: A Critical Examination of the Foundation of Analysis, Hermann Weyl. Classic of 20th-century foundational research deals with the conceptual problem posed by the continuum. 156pp. 5⅜ x 8½. 67982-9 Pa. $6.95

ESSAYS ON THE THEORY OF NUMBERS, Richard Dedekind. Two classic essays by great German mathematician: on the theory of irrational numbers; and on transfinite numbers and properties of natural numbers. 115pp. 5⅜ x 8½.
21010-3 Pa. $5.95

THE FUNCTIONS OF MATHEMATICAL PHYSICS, Harry Hochstadt. Comprehensive treatment of orthogonal polynomials, hypergeometric functions, Hill's equation, much more. Bibliography. Index. 322pp. 5⅜ x 8½. 65214-9 Pa. $9.95

NUMBER THEORY AND ITS HISTORY, Oystein Ore. Unusually clear, accessible introduction covers counting, properties of numbers, prime numbers, much more. Bibliography. 380pp. 5⅜ x 8½. 65620-9 Pa. $10.95

THE VARIATIONAL PRINCIPLES OF MECHANICS, Cornelius Lanczos. Graduate level coverage of calculus of variations, equations of motion, relativistic mechanics, more. First inexpensive paperbound edition of classic treatise. Index. Bibliography. 418pp. 5⅜ x 8½. 65067-7 Pa. $12.95

MATHEMATICAL TABLES AND FORMULAS, Robert D. Carmichael and Edwin R. Smith. Logarithms, sines, tangents, trig functions, powers, roots, reciprocals, exponential and hyperbolic functions, formulas and theorems. 269pp. 5⅜ x 8½.
60111-0 Pa. $6.95

THEORETICAL PHYSICS, Georg Joos, with Ira M. Freeman. Classic overview covers essential math, mechanics, electromagnetic theory, thermodynamics, quantum mechanics, nuclear physics, other topics. First paperback edition. xxiii + 885pp. 5⅜ x 8½. 65227-0 Pa. $21.95

ORDINARY DIFFERENTIAL EQUATIONS, Morris Tenenbaum and Harry Pollard. Exhaustive survey of ordinary differential equations for undergraduates in mathematics, engineering, science. Thorough analysis of theorems. Diagrams. Bibliography. Index. 818pp. 5⅜ x 8½. 64940-7 Pa. $18.95

STATISTICAL MECHANICS: Principles and Applications, Terrell L. Hill. Standard text covers fundamentals of statistical mechanics, applications to fluctuation theory, imperfect gases, distribution functions, more. 448pp. 5⅜ x 8½. 65390-0 Pa. $11.95

ORDINARY DIFFERENTIAL EQUATIONS AND STABILITY THEORY: An Introduction, David A. Sánchez. Brief, modern treatment. Linear equation, stability theory for autonomous and nonautonomous systems, etc. 164pp. 5⅜ x 8¼. 63828-6 Pa. $6.95

THIRTY YEARS THAT SHOOK PHYSICS: The Story of Quantum Theory, George Gamow. Lucid, accessible introduction to influential theory of energy and matter. Careful explanations of Dirac's anti-particles, Bohr's model of the atom, much more. 12 plates. Numerous drawings. 240pp. 5⅜ x 8½. 24895-X Pa. $7.95

THEORY OF MATRICES, Sam Perlis. Outstanding text covering rank, nonsingularity and inverses in connection with the development of canonical matrices under the relation of equivalence, and without the intervention of determinants. Includes exercises. 237pp. 5⅜ x 8½. 66810-X Pa. $8.95

GREAT EXPERIMENTS IN PHYSICS: Firsthand Accounts from Galileo to Einstein, edited by Morris H. Shamos. 25 crucial discoveries: Newton's laws of motion, Chadwick's study of the neutron, Hertz on electromagnetic waves, more. Original accounts clearly annotated. 370pp. 5⅜ x 8½. 25346-5 Pa. $10.95

INTRODUCTION TO PARTIAL DIFFERENTIAL EQUATIONS WITH APPLICATIONS, E.C. Zachmanoglou and Dale W. Thoe. Essentials of partial differential equations applied to common problems in engineering and the physical sciences. Problems and answers. 416pp. 5⅜ x 8½. 65251-3 Pa. $11.95

BURNHAM'S CELESTIAL HANDBOOK, Robert Burnham, Jr. Thorough guide to the stars beyond our solar system. Exhaustive treatment. Alphabetical by constellation: Andromeda to Cetus in Vol. 1; Chamaeleon to Orion in Vol. 2; and Pavo to Vulpecula in Vol. 3. Hundreds of illustrations. Index in Vol. 3. 2,000pp. 6⅛ x 9¼. 23567-X, 23568-8, 23673-0 Pa., Three-vol. set $44.85

CHEMICAL MAGIC, Leonard A. Ford. Second Edition, Revised by E. Winston Grundmeier. Over 100 unusual stunts demonstrating cold fire, dust explosions, much more. Text explains scientific principles and stresses safety precautions. 128pp. 5⅜ x 8½. 67628-5 Pa. $5.95

AMATEUR ASTRONOMER'S HANDBOOK, J.B. Sidgwick. Timeless, comprehensive coverage of telescopes, mirrors, lenses, mountings, telescope drives, micrometers, spectroscopes, more. 189 illustrations. 576pp. 5⅝ x 8¼. (Available in U.S. only) 24034-7 Pa. $11.95

GEOMETRY OF COMPLEX NUMBERS, Hans Schwerdtfeger. Illuminating, widely praised book on analytic geometry of circles, the Moebius transformation, and two-dimensional non-Euclidean geometries. 200pp. 5⅝ x 8¼. 63830-8 Pa. $8.95

MECHANICS, J.P. Den Hartog. A classic introductory text or refresher. Hundreds of applications and design problems illuminate fundamentals of trusses, loaded beams and cables, etc. 334 answered problems. 462pp. 5⅜ x 8½. 60754-2 Pa. $11.95

TOPOLOGY, John G. Hocking and Gail S. Young. Superb one-year course in classical topology. Topological spaces and functions, point-set topology, much more. Examples and problems. Bibliography. Index. 384pp. 5⅜ x 8¼. 65676-4 Pa. $10.95

STRENGTH OF MATERIALS, J.P. Den Hartog. Full, clear treatment of basic material (tension, torsion, bending, etc.) plus advanced material on engineering methods, applications. 350 answered problems. 323pp. 5⅜ x 8½. 60755-0 Pa. $9.95

ELEMENTARY CONCEPTS OF TOPOLOGY, Paul Alexandroff. Elegant, intuitive approach to topology from set-theoretic topology to Betti groups; how concepts of topology are useful in math and physics. 25 figures. 57pp. 5⅜ x 8½. 60747-X Pa. $3.95

ADVANCED STRENGTH OF MATERIALS, J.P. Den Hartog. Superbly written advanced text covers torsion, rotating disks, membrane stresses in shells, much more. Many problems and answers. 388pp. 5⅜ x 8½. 65407-9 Pa. $10.95

COMPUTABILITY AND UNSOLVABILITY, Martin Davis. Classic graduate-level introduction to theory of computability, usually referred to as theory of recurrent functions. New preface and appendix. 288pp. 5⅜ x 8½. 61471-9 Pa. $8.95

GENERAL CHEMISTRY, Linus Pauling. Revised 3rd edition of classic first-year text by Nobel laureate. Atomic and molecular structure, quantum mechanics, statistical mechanics, thermodynamics correlated with descriptive chemistry. Problems. 992pp. 5⅜ x 8½. 65622-5 Pa. $19.95

AN INTRODUCTION TO MATRICES, SETS AND GROUPS FOR SCIENCE STUDENTS, G. Stephenson. Concise, readable text introduces sets, groups, and most importantly, matrices to undergraduate students of physics, chemistry, and engineering. Problems. 164pp. 5⅜ x 8½. 65077-4 Pa. $7.95

THE HISTORICAL BACKGROUND OF CHEMISTRY, Henry M. Leicester. Evolution of ideas, not individual biography. Concentrates on formulation of a coherent set of chemical laws. 260pp. 5⅜ x 8½. 61053-5 Pa. $8.95

THE PHILOSOPHY OF MATHEMATICS: An Introductory Essay, Stephan Körner. Surveys the views of Plato, Aristotle, Leibniz & Kant concerning propositions and theories of applied and pure mathematics. Introduction. Two appendices. Index. 198pp. 5⅜ x 8½. 25048-2 Pa. $8.95

THE DEVELOPMENT OF MODERN CHEMISTRY, Aaron J. Ihde. Authoritative history of chemistry from ancient Greek theory to 20th-century innovation. Covers major chemists and their discoveries. 209 illustrations. 14 tables. Bibliographies. Indices. Appendices. 851pp. 5⅜ x 8½. 64235-6 Pa. $18.95

CATALYSIS IN CHEMISTRY AND ENZYMOLOGY, William P. Jencks. Exceptionally clear coverage of mechanisms for catalysis, forces in aqueous solution, carbonyl- and acyl-group reactions, practical kinetics, more. 864pp. 5⅜ x 8½.
65460-5 Pa. $19.95

PROBABILITY: An Introduction, Samuel Goldberg. Excellent basic text covers set theory, probability theory for finite sample spaces, binomial theorem, much more. 360 problems. Bibliographies. 322pp. 5⅜ x 8½. 65252-1 Pa. $10.95

LIGHTNING, Martin A. Uman. Revised, updated edition of classic work on the physics of lightning. Phenomena, terminology, measurement, photography, spectroscopy, thunder, more. Reviews recent research. Bibliography. Indices. 320pp. 5⅜ x 8¼. 64575-4 Pa. $8.95

PROBABILITY THEORY: A Concise Course, Y.A. Rozanov. Highly readable, self-contained introduction covers combination of events, dependent events, Bernoulli trials, etc. Translation by Richard Silverman. 148pp. 5⅜ x 8¼. 63544-9 Pa. $7.95

AN INTRODUCTION TO HAMILTONIAN OPTICS, H. A. Buchdahl. Detailed account of the Hamiltonian treatment of aberration theory in geometrical optics. Many classes of optical systems defined in terms of the symmetries they possess. Problems with detailed solutions. 1970 edition. xv + 360pp. 5⅜ x 8½.
67597-1 Pa. $10.95

STATISTICS MANUAL, Edwin L. Crow, et al. Comprehensive, practical collection of classical and modern methods prepared by U.S. Naval Ordnance Test Station. Stress on use. Basics of statistics assumed. 288pp. 5⅜ x 8½. 60599-X Pa. $7.95

DICTIONARY/OUTLINE OF BASIC STATISTICS, John E. Freund and Frank J. Williams. A clear concise dictionary of over 1,000 statistical terms and an outline of statistical formulas covering probability, nonparametric tests, much more. 208pp. 5⅜ x 8½. 66796-0 Pa. $7.95

STATISTICAL METHOD FROM THE VIEWPOINT OF QUALITY CONTROL, Walter A. Shewhart. Important text explains regulation of variables, uses of statistical control to achieve quality control in industry, agriculture, other areas. 192pp. 5⅜ x 8½. 65232-7 Pa. $7.95

METHODS OF THERMODYNAMICS, Howard Reiss. Outstanding text focuses on physical technique of thermodynamics, typical problem areas of understanding, and significance and use of thermodynamic potential. 1965 edition. 238pp. 5⅜ x 8½.
69445-3 Pa. $8.95

STATISTICAL ADJUSTMENT OF DATA, W. Edwards Deming. Introduction to basic concepts of statistics, curve fitting, least squares solution, conditions without parameter, conditions containing parameters. 26 exercises worked out. 271pp. 5⅜ x 8½.
64685-8 Pa. $9.95

TENSOR CALCULUS, J.L. Synge and A. Schild. Widely used introductory text covers spaces and tensors, basic operations in Riemannian space, non-Riemannian spaces, etc. 324pp. 5⅜ x 8¼. 63612-7 Pa. $9.95